Science in the American Southwest

Science

in the American Southwest

A Topical History

GEORGE E. WEBB

The University of Arizona Press
Tucson

The University of Arizona Press
© 2002 The Arizona Board of Regents
First printing
All rights reserved
⊚ This book is printed on acid-free, archival-quality paper.
Manufactured in the United States of America
07 06 05 04 03 02   6 5 4 3 2 1

Library of Congress Cataloging-in-Publication Data
Webb, George Ernest.
Science in the American Southwest : a topical history / George E. Webb.
p. cm.
Includes bibliographical references and index.
ISBN 0-8165-2188-3 (cloth : alk. paper)
1. Research—Arizona—History—20th century. 2. Research—New Mexico—History—
20th century. 3. Research—Texas—History—20th century. I. Title.
Q180.U5 W37 2002
507'.2079—dc21
2001008028

British Library Cataloging-in-Publication Data
A catalog record for this book is available from the British Library.

Frontispiece: Two of the major telescopes on Kitt Peak. The dome on the right houses the
158-inch Mayall Telescope, completed in 1973. On the left is the Steward Observatory
90-inch reflector, completed in 1969. (Photograph by the author)

In memory of
Herman E. Bateman
and
Gerald E. Thompson

Valued colleagues and friends

Contents

Part III The Foundations of Modern Science

Figures

Tables

Acknowledgments

Scholarly projects are invariably collaborative efforts. They are based on the previous work of others, as well as on the author's interaction with colleagues. The bibliography of this volume indicates the great debt I owe to earlier investigators, but numerous colleagues have also contributed significantly to the completion of this project. Preliminary research was conducted during a non-instructional assignment at Tennessee Technological University during the spring of 1998. Of particular value during this period was a research trip to consult various archival and other sources in Tucson, made possible through the actions of William Brinker, history department chairman. Department secretary Lois Clinton took care of the administrative details of this process with her usual aplomb and good humor. Librarians and archivists in the Southwest have provided assistance in many ways and for many years. Especially helpful have been those at the Arizona Historical Society, the University of Arizona Library Special Collections, the Los Alamos Historical Museum, the State Records Center and Archives in Santa Fe, and the libraries at the University of New Mexico and New Mexico State University. Special thanks go to Rebecca Collinsworth of the Los Alamos Historical Museum and to Maria Schuchardt of the Space Imagery Center of the University of Arizona, both of whom provided valuable photographs from their collections. The following journals have graciously given permission to use material previously published in their periodicals: the *Journal of Arizona History*, the *New Mexico Historical Review*, the *Journal of the Southwest*, *The Historian*, *Isis*, and the *Bulletin for the History of Chemistry*.

Several individuals have made contributions during the course of this project. John Lankford has offered good advice and much encouragement over the years, as have Roger Nichols, Michael Allen, Robert E. Cray, Bryant Bannister, and the late Gerald Thompson. David Strauss and Ferenc M. Szasz carefully read the completed manuscript and made several excellent suggestions that improved the finished product considerably. Historians rely heavily on interlibrary loan personnel to secure material not otherwise available. At Tennessee Tech, Linda Mulder consistently retrieved esoteric material from libraries throughout the United States. This project would not have been possible without her assistance. Working with the University of Arizona Press has again been a pleasure, thanks to the efforts of Patti Hartmann, Al Schroder, and Rose Byrne. Finally, to my mother, I offer sincere gratitude for five decades of constant and unfailing support.

Introduction

As the twentieth century merged into the twenty-first, observers of American science often cast their gaze toward the Southwest. As it had for more than five decades, Los Alamos continued to attract much interest, although security lapses and destructive fires attracted as much attention as the fundamental research pursued by the laboratory's scientists. Anthropologists and archaeologists continued to explore the region's ruins, adding significantly to knowledge of the ancient inhabitants of the Southwest. Astronomy remained a significant specialty among the region's scientists, as noted by reports of dramatic discoveries and new instruments. In addition to observatories funded by governments and private sources, the Southwest also hosted major universities that supported programs in astronomy and space science. These programs served as the base for the region's involvement in the American space effort, as shown by the many NASA missions that included large numbers of researchers from southwestern institutions. To augment these examples of the region's contributions to modern science, the well-established universities of the Southwest supported large science departments in many fields, providing the needed institutional base to pursue research and education in a wide variety of disciplines.

It is, thus, somewhat curious that relatively little attention has been paid to the historical development of science in the American Southwest. Individual studies have appeared over the years, but neither historians of the American West nor historians of science have devoted their energies to the topic. For a brief period in the 1970s and early 1980s, new trends in western history held promise to correct this oversight. As exemplified by the pioneering work of Gerald D. Nash, historians began to examine in detail the twentieth-century West, focusing on the region's post-frontier development. Especially in the years following World War II, science played an increasingly important role throughout the United States, a phenomenon reflected in the West. The establishment and growth of Los Alamos was an obvious example of scientific development, but Nash and his colleagues also stressed such institutional developments as the growth of universities and laboratories and their impact on establishing a viable scientific community. Admittedly, much of the attention paid to these developments focused on California and its emergence as one of the scientific centers of the United States, but historians also examined developments elsewhere in the region.[1]

The potential for a more complete view of science in the West remained unrealized, however, as the focus of the discipline changed during the 1980s to embrace what became known as the New Western History. Science was not completely ignored in this new vision of the western past, but its role in the region's history was dramatically redefined. Many historians focused their attention on environmental topics, reflecting contemporary concerns. These historians found much of interest in the examination of public policy and the role scientists played in setting such policy. The science and scientists involved in such developments, however, were less important than the impact of such policy on the environment. The atomic research pursued in the region, for example, was examined not as a representative case of institutional changes and their impact on knowledge in physics and chemistry, but as an example of environmental degradation and the negative impact of humans on the region.[2]

The negative focus of much of the New Western History, noted by various critics, led historians away from an examination of science in other ways. These scholars rejected the positive tone of earlier histories of the West, denigrating the work of Frederick Jackson Turner, Ray Allen Billington, and others as "celebratory" of the "conquest" of the American

West. Rejecting concepts of "progressive" history generally, these historians had little use for topics that displayed advances over time. The development of science must be understood as a complex phenomenon, to be certain, but it remains characterized by the cumulative addition of knowledge that is, in a very real sense, progressive. Science is also very much an "elite" activity, in the sense that relatively few individuals practice any of the various scientific disciplines. By ability and education, scientists remain somewhat apart from the rest of society. This characteristic also troubled the new western historians, who increasingly wanted to study "the ordinary lives of ordinary people," reflecting contemporary trends in social history. Despite their calls for more "inclusive" studies, the new western historians generally failed to include science or scientists in their new vision of the American West.[3]

Historians of science have been no more interested in the Southwest as a research field. Although the study of American science has become the leading subfield in the discipline over the past few decades, historians have concentrated on developments in the eastern half of the nation, with occasional examinations of events in California. Such a focus is understandable, as these areas contained large numbers of scientists and scientific institutions. Especially in the twentieth century, however, other areas of the nation also participated in the growth of science, providing new data and insights in many disciplines. The American Southwest was an excellent source of such contributions throughout the late nineteenth and twentieth centuries, even before the region's institutional base was established. And yet, with the exception of several studies of Los Alamos during the World War II and early Cold War periods, historians of science have shown relatively little interest in the region. Even in the Los Alamos studies, the connection with such leading Manhattan Project institutions as the University of California and the University of Chicago often remained of greater interest.[4]

Nonetheless, the history of the region's science has been explored by a handful of scholars who have not been constrained by disciplinary fashions. Because of its long importance in the region's science, astronomy has attracted much attention. Biographies of such astronomers as Percival Lowell and A. E. Douglass, as well as studies of various observatories, have added much to our knowledge of both the scientific and institutional factors involved in the region's development.[5] Anthropologists and archaeologists have surveyed the historical developments in their disciplines, providing an excellent overview of the evolution of one of the leading scientific disciplines in the Southwest.[6] More specialized studies of other fields have also occasionally appeared, although a notable lack of institutional histories has restricted efforts to create a foundation for integrating individual studies into a coherent whole.[7]

The purpose of this volume, therefore, is to call attention to the central role science has played in the development of the American Southwest. Largely ignored, science has contributed as much to the emergence of the modern Southwest as the political, economic, and social developments that have long been the concern of historians. Rather than attempt to present an overview of the development of science in the region, however, I examine several topics in the history of science in the Southwest, each of which points toward broader topics for future analysis. The individuals and institutions involved in these pursuits display numerous characteristics of science in the region and provide a guide to the various ways in which science was established and grew. Such studies are an important first step toward a broader knowledge and appreciation of science in the American Southwest as well as toward a more complete vision of the intellectual and cultural growth of the region.

Definitions and Decisions

Neither *Southwest* nor *science* are unambiguous terms. Historians of the American West have long struggled to define the larger region, rarely agreeing on the boundaries involved. The various subregions of the West have been no easier to delimit. *Southwest* has often been used to describe the lands between east Texas and the Pacific coast, and between the Mexican boundary and the northern borders of Nevada, Utah, and Colorado. Although this definition displays a certain cartographic logic, it obscures important differences. The California coastline has little in common with the southern Arizona deserts, while northern Utah remains quite different from the Rio Grande valley in New Mexico. In an effort to determine and describe the unique aspects of science in the American Southwest, therefore, a more narrowly defined region is necessary. The present study has thus been restricted to the area including the current states of Arizona and New Mexico, as well as that part of Texas west of the Pecos River. So defined, the region shares many geographical characteristics that have long proved of interest to scientists in various disciplines. Equally important, the institutional, cultural, and political development of the region displays a degree of continuity that has had an important bearing on the growth of science in the Southwest.

Similarly difficult is the decision concerning which aspects of science to include in this study. In order to emphasize the diversity of scientific development in the region, a broad coverage is necessary, as is a focus on topics not previously explored in depth. The topical approach is clearly of value in such a project, but the topics to be studied must be selected with care. The pursuit of astronomical knowledge, the application of science to mining and agriculture, the growth of a regional scientific community, and the establishment of "big science" programs all provide insight into the importance of science to the American Southwest. The public response to scientific developments and concepts, as shown by the continuing controversy over the teaching of evolution, reveals the broader societal component of modern science and further illustrates the important place held by science in the region.

The topics that are discussed not only indicate the wide variety of scientific developments in the Southwest, but also stress four themes that appear to characterize and unite these developments. From the earliest expressions of scientific interest in the region, the concept of the Southwest as a "natural laboratory" guided activity. Explorers noted the unique environmental characteristics caused by the region's aridity and were intrigued with the numerous examples of previously unknown plants and animals. As better-trained naturalists began to arrive in the late nineteenth century, they described these examples of flora and fauna for scientific periodicals and frequently sent specimens to museums in the East. The region's role as a natural laboratory had an impact on several scientific disciplines. Archaeologists and anthropologists certainly viewed the Southwest as a laboratory for their disciplines, while astronomers increasingly focused their attention on the region because of its climatological superiority for observational efforts.

The recognition of the Southwest as a source of new data and opportunities led to another theme in the development of the region's science. Until well after World War II, the Southwest remained a "colony" of the national scientific establishment. During the late nineteenth and early twentieth centuries, eastern scientists examined the region as a source of new knowledge, returning to their home institutions to process this information. As the institutional base of southwestern science grew in the first half of the twentieth century, the connection with eastern institutions (and, to a lesser extent, California facilities) remained important. Professional encouragement and institutional funding were both crucial to the growth of new laboratories and observatories. Such support was often augmented by scientists who traveled to the Southwest to oversee research projects. Over time, a growing number of scientists relocated from their previous positions to the new facilities in the Southwest, providing a more permanent, if still somewhat colonial, presence.

Institutional growth also characterized the development of science in the American Southwest and represents another theme. Among the most dramatic examples of this theme was the emergence of astronomy in the region. Beginning with the Lowell Observatory in 1894, astronomy became increasingly important in the region's scientific expansion. By the 1930s, the Steward Observatory in southern Arizona and the McDonald Observatory in west Texas had been completed and had joined the Lowell facility at the forefront of astronomical research. Beginning in the late 1940s, the role played by southwestern scientists in the American space program led to significant institutional growth as well. The expansion of the White Sands facility during the early years of the space program was followed by the establishment of such institutions as the Lunar and Planetary Laboratory at the University of Arizona. Although these and other facilities were funded by the federal government, they were often associated with educational institutions. The Southwest's colleges and universities remained an important part of the region's scientific infrastructure throughout the twentieth century but became even more significant following World War II. As several of the region's universities became recognized as research institutions during the 1970s and 1980s, their role in the pursuit of science grew dramatically and was usually accompanied by increased outside funding from both private and public sources.

A final theme that characterizes the development of science in the American Southwest is the role played by the federal government. Federal funding has been central to American science since World War II, a phenomenon in which the Southwest has shared fully. Astronomy (as indicated by the establishment of Kitt Peak National Observatory) and the space sciences have been the most noteworthy recipients of this largesse, but few research projects were without at least some federal funding. As part of the Southwest's existence as a natural laboratory and a scientific colony, however, the region participated in federally sponsored research even before the changes brought about by World War II. Scientific research was supported by the U.S. Department of Agriculture through the experiment stations of Arizona and New Mexico, beginning in the territorial period. The department also supported research in forest and range management. Various other scientific projects also gained support during the New Deal, as the federal government distributed money to universities, museums, and other institutions as part of its relief efforts. In the Southwest, archaeological investigations benefited greatly from such funding, but numerous projects on university campuses also secured research funds. The New Deal expansion of existing federal programs, especially those of the Department of Agriculture, also contributed to the growth of science in the American Southwest.

These four themes both characterize the growth of science in the American Southwest and serve as the unifying concepts in the historical examination of the topic. The region's scientific development has played a crucial role in its emergence into national prominence, but that role remains inadequately studied. The examination of several examples of the role of science in the American Southwest thus provides a more complete perspective on the region's development and offers valuable insight concerning the region's place in the national context.

Part I The Establishment of Science

Introduction

Until well into the twentieth century, the American Southwest was of greatest interest to scientists as a "natural laboratory." Explorers provided descriptions of the region that attracted the interest of naturalists in Europe and, later, the United States. As trained scientists became increasingly involved in the exploration of the Southwest, they furnished colleagues with significant information and forwarded specimens of previously unknown plants and animals that swelled the collections of museums. Well-preserved archaeological ruins were studied in an effort to comprehend the region's earliest cultures, as were paleontological deposits that expanded scientists' understanding of the prehistoric world. Even after an institutional foundation had been laid in the late nineteenth and early twentieth century, the region's natural characteristics continued to define its contributions to scientific knowledge.

The native inhabitants of the Southwest were also concerned with the natural characteristics of the region, albeit for utilitarian purposes. Their knowledge of indigenous plants provided treatment for various maladies, while their recognition of the demands of the arid climate led to dry farming techniques, including irrigation, that provided a reliable food supply. By making careful and continuous observations of the stars and planets, native peoples developed calendars to guide planting activity. While none of these activities represented "mainstream" science, especially as it was practiced by Europeans after the sixteenth century, awareness of the region's unique characteristics shaped the earliest development of the American Southwest.

Spanish exploration of the Southwest was also guided by utilitarian concerns. These explorers focused on gathering useful information to support Spain's imperial ambitions. Observations of plants, animals, and geography were recorded, but Spanish policy restricted the dissemination of this knowledge. It thus remained buried in archives for generations. Military and civilian explorers sent to the Southwest by the United States government in the mid-nineteenth century conducted more systematic surveys of the region. Military needs, settlement, and transportation remained the focus of this reconnaissance, but the expeditions included scientists in field parties and were advised by leading naturalists from eastern universities and museums.

Following the Civil War, extensive examinations of the Southwest became increasingly common. These later surveys were financed by the federal government and by private support from museums, universities, and wealthy patrons. A wide variety of endeavors characterized these expeditions, but the natural history focus remained primary. Botanists and zoologists carefully surveyed the region and sent large collections of specimens to museums in Boston, Philadelphia, New York, and Washington, D.C. The observations of field geologists not only added significant data concerning the region's physical structure, but also provided theoretically minded colleagues with crucial evidence for uniformitarian explanations based on the role of erosion.

Among the more dramatic results from these efforts were important revelations in paleontology. During the 1870s and 1880s, noted paleontologist Edward Drinker Cope analyzed recently discovered fossil deposits in New Mexico. The San Ildefonso beds near Santa Fe revealed large mammals such as mastodons, camels, horses, and horned deer, but the Puerco formation in the northwestern portion of the territory proved even more significant. Some seventy million years old, this formation provided Cope with nearly a hundred species of small animals from the beginning of the Age of Mammals. These fossils added significantly to scientists' understanding of the emergence of mammals and established Cope as one of the leading paleontologists of the nation. He presented the results of his research at various professional conferences and published accounts of his efforts in such periodicals as the *Proceedings* of the American Philosophical Society and *American Naturalist*.[1]

Beginning in the 1890s, the American Museum of Natural History sent several expeditions to the area, providing even more detailed information concerning early mammalian fossils. In addition, museum paleontologists examined the strata beneath those explored by Cope and others, finding many dinosaur fossils from the late Cretaceous period. These discoveries from the end of the Age of Dinosaurs, especially when integrated with mammalian fossils from the same geographical area, allowed scientists to analyze the dramatic transition from the Mesozoic to the Cenozoic era.[2]

Applied science also played an important role in the development of a scientific presence in the American Southwest. Mining activity had long been a central aspect of the region's economic development, but the transition from placer to hard-rock mining after the Civil War brought with it the need for more specialized expertise. Mining engineers superintended the design and construction of mines and mills and frequently guided the operation of such facilities. Chemists, geologists, and engineers also served as consultants to the Southwest's mineral industry, advising potential investors concerning the value of mining properties. Scientists such as Yale chemist Benjamin Silliman Jr. were key figures in the dramatic expansion of the industry in the late nineteenth century and established another role for scientists in the region.[3]

Several individuals contributed in this fashion to the growth of mining in the Southwest. Among the more noteworthy of these scientists were Henry and Alexis Janin. Henry examined mineral deposits along the Colorado River in the mid-1860s, traveling by mule-drawn ambulance and accompanied by a valet/servant who took care of his wardrobe and portable wine cellar. Younger brother Alexis traveled in central Arizona in the late 1870s, advising clients about numerous mining properties. During the same decade, James Douglas began his momentous career in the Arizona copper industry as a consultant to investors involved with the Copper Queen Mine in southeastern Arizona and the later Phelps Dodge efforts in the Jerome area. Douglas quickly emerged as an outstanding example of the engineer/executive and became quite successful and wealthy as a result of his efforts. Another important figure in the mineral industry was William P. Blake, who examined mining properties throughout the Southwest. After forty years of such activity, Blake became director of the Arizona School of Mines in 1895.[4]

Blake's position with the mining school in Tucson suggests another aspect of the growing importance of science. During the late 1880s, Arizona and New Mexico created their territorial university systems, at least in part to provide technical and educational support for the region's mining industry. Arizona established a School of Mines within its university, while the New Mexico legislature determined that a separate mining school in Socorro would be more appropriate. In each case, the institutions provided employment for chemists, geologists, and engineers, who contributed their expertise to the mining industry. These schools were complemented in 1914 with the establishment of the Texas State School of Mines and Metallurgy in El Paso.[5]

Agriculture served as the other foundation for the region's economy and was also characterized by an increased concern with scientific considerations. Agricultural experiment stations, funded by the U.S. Department of Agriculture, were an especially important component of the territorial university systems. The University of Arizona established an agricultural program primarily to secure sufficient federal funds to build and operate the university. The New Mexico College of Agriculture and Mechanic Arts was founded in Las Cruces in much the same fashion. Not only did these stations provide financial support (and, on occasion, teaching faculty) for their host institutions, but they also brought chemists and biologists into the region to improve the efficiency and profitability of southwestern agriculture. As research activity became an increasingly important part of the federal agriculture program, experiment stations served as the focus of efforts to expand the region's agricultural endeavors.[6]

Among the more intriguing research programs conducted by experiment station scientists were the efforts to establish dates as a marketable product. Beginning in the early 1890s, the Arizona experiment station sought to capitalize on the territory's climate by finding a desert crop that would be commercially viable. These initial efforts led to a cooperative arrangement in 1898 between the station and the Bureau of Plant Industry to import various date palm offshoots for research by Arizona scientists. Over the next decade, some 400 plants were cultivated at branch stations in Tempe and Yuma in an attempt to determine which varieties would be most successful.[7] Between 1905 and 1907, station staff examined the *Parlatoria* scale, a dangerous insect that threatened date production. Eradication was eventually accomplished through the application of gasoline and a blowtorch, after which the date palm grew without scale infestation. Chemists and botanists conducted numerous experiments in their efforts to ripen dates artificially, using chemicals and heat. This program was designed to simplify shipment of the fruit, as dates could be ripened at their destination to avoid spoilage. Although the Arizona date industry never achieved the level of success enthusiasts had anticipated, the work of experiment station scientists was applied to efforts in the Coachella Valley of California, which soon became the center of North American date production.[8]

The U.S. Department of Agriculture also supported other research projects in the Southwest during the early twentieth century. In 1907, for example, the Bureau of Plant Industry established its Cotton Research Center near Sacaton, Arizona. This field laboratory on the Pima Indian Reservation soon perfected the long staple variety known as pima cotton, which entered commercial production in 1912. The Forest Service founded the nation's first forest experiment station near Flagstaff in 1908, implementing suggestions made by Gifford Pinchot and other conservationists during the previous decade. The success of this facility in improving forestry management led to the establishment of similar stations in Colorado, Idaho, Washington, California, and Utah, again showing the importance of applied science to agriculture. Many of the scientists in the American Southwest at this time were associated with institutions pursuing such activities.[9]

The climatic characteristics of the American Southwest often provided the base for significant scientific contributions. As telescopes increased in size during the late nineteenth century, astronomers recognized the importance of clear and steady air for the most efficient use of these instruments. The Southwest attracted the attention of many astronomers for this reason. When Percival Lowell decided to devote a portion of his fortune to the erection of an observatory to study the planet Mars, he accepted the advice of friends at the Harvard College Observatory and focused his attention on Arizona and New Mexico. Lowell sent his assistant, A. E. Douglass, to survey potential sites during the spring of 1894. Seeking superior atmospheric conditions, the young astronomer surveyed several areas in Arizona before Lowell instructed him to restrict his survey to the Flagstaff region. Preliminary results convinced Lowell that atmospheric conditions were suitable, but he was equally insistent that the facility be completed in time to take advantage of Mars's close approach to Earth during the summer. Douglass located an ideal site on a mesa west of town later known as Mars Hill and quickly began construction of the Lowell Observatory. Despite Lowell's questionable reputation among astronomers for his imaginative views of an ancient Martian civilization, the observatory soon became an important part of the astronomical community. Especially after the death of its founder in 1916, the Flagstaff facility contributed significantly to both galactic and planetary astronomy.[10]

The climate of the arid Southwest served another scientific discipline as well. Numerous ancient habitations had been well preserved by desert conditions, providing archaeologists and anthropologists with valuable research material. Early students of the region's ancient civilizations began their investigations in the late nineteenth century, locating and excavating ruins throughout the Southwest. Such noted figures as Adolph Bandelier, Frank Hamilton Cushing, and Jesse Walter Fewkes provided colleagues in eastern museums and universities with artifacts and descriptions that yielded new perspectives on North American prehistory. The region's supply of ruins also attracted institutional interest, leading the Archaeological Institute of America to establish a research facility in Santa Fe. Directed by Edgar Lee Hewett, this facility soon became known as the School of American Archaeology. Hewett was also instrumental in establishing the Museum of New Mexico to expand research in the region and to disseminate the results of various investigations. Early in the twentieth century, the Southwest's ruins came under more intensive analysis through the work of Alfred Vincent Kidder, Earl H. Morris, and Frederick Webb Hodge. These and other archaeologists further established the region as the source for significant new knowledge and as a training ground for future generations of scholars.[11]

The native vegetation of the desert continued to be of great interest as the twentieth century dawned, although botanists were now concerned with more than the collection and cataloging of new specimens. The extreme conditions of the harsh desert climate revealed much about the impact of environmental factors on plant growth. In 1902, therefore, the Carnegie Institution of Washington decided to establish a laboratory to investigate desert plants. By the end of the year, the Carnegie directors had authorized \$8,000 for the facility and had determined that Tucson would be the best location for the Carnegie Desert Botanical Laboratory. The highly regarded botanist Daniel T. MacDougal became director three years later and coordinated the research program until his retirement in the late 1920s. Although the laboratory closed in 1940, its studies of desert vegetation added significantly to botanical knowledge.[12]

The impact of the desert climate on vegetation was also central to one of the most dramatic contributions of Southwest science in the early twentieth century. The development of dendrochronology would have far-reaching implications for climatology and archaeology, but its origins lay elsewhere. As were many of his colleagues, astronomer A. E. Douglass was intrigued with the possible connection between solar activity (especially sunspots) and terrestrial climate. Douglass recognized that precipitation was the prime determinant of tree growth in the desert Southwest and concluded that if solar activity had an impact on precipitation, this impact would be reflected in the annual growth rings of trees. Tree rings could thus provide a record of past climate and, if they accurately recorded past solar activity, an indirect catalog of sunspots. Although Douglass failed to establish a direct connection between solar activity and terrestrial climate, his construction of centuries-long chronologies enabled the accurate dating of archaeological ruins and provided an extended record of past climatic conditions.[13]

As the twentieth century opened, scientists in the American Southwest were expanding knowledge in several disciplines. Although the institutional base of science in the region had been tentatively established, much of these scientists' work remained focused on the unique characteristics of the Southwest. Their research would lead to greater knowledge of natural history and would often have practical benefits for agriculture and mining. The region's climate attracted the attention of astronomers and botanists, who recorded and explained observed phenomena. The natural laboratory of the American Southwest had thus not only added significantly to the catalog of North American flora and fauna, but it had also laid the foundation for future developments that would fundamentally alter scientists' understanding of the natural world.

Chapter 1 The Desert Revealed

The Origins of Southwestern Science

For three and one-half centuries, the American Southwest served scientists primarily as a source of new information. Previously unknown species of plants and animals, as well as the landscape itself, made dramatic impressions on observers from diverse backgrounds and with various interests. As they journeyed through the region from the sixteenth into the eighteenth century, Spanish explorers noted the geological and biological resources of the Southwest, while making observations that would improve existing maps. Although official policy restricted the dissemination of this information, the collection of knowledge concerning the region characterized Spanish activity in the New World.

By the mid-nineteenth century, the American Southwest had assumed an important place in the perspective of the United States. Controlling most of the region following its war with Mexico, the United States became increasingly intrigued with the resources of the Southwest and dispatched numerous explorers to survey the territory. These surveys were almost exclusively devoted to practical concerns, but their contributions to scientific knowledge were considerable. Specimens collected in the region were sent to scientists at the leading institutions in the East for classification, building the collections of museums in a dramatic fashion.

The Smithsonian Institution, in fact, became the nation's leading natural history museum primarily as a result of the western surveys of the mid-nineteenth century. Cooperative ventures of military personnel and civilian scientists, these surveys provided a comprehensive catalog of the natural resources and characteristics of the American Southwest.

Following the Civil War, however, the scientific exploration of the Southwest took on a different perspective. Increasingly, the naturalist tradition of careful observation, description, and analysis of specimens characterized the work of those intrigued with the region's flora and fauna. Throughout the second half of the nineteenth century, and well into the twentieth, the Southwest witnessed an ever-growing number of visitors who recorded their observations of the region. Many of these naturalists were staff members of eastern institutions, while a significant number financed their efforts through private funds or by selling specimens to museums and collectors. Their observations were published in scientific periodicals, providing a new awareness of the region's environmental characteristics to scientists who would never visit the Southwest.

The Natives

When Europeans entered the Southwest in the sixteenth century, they came upon existing cultures which they found of great interest. As these new arrivals explored the region, they discovered that these cultures had developed societies uniquely suited to the harsh environment. Equally intriguing, ruins of earlier civilizations suggested that the Southwest had been inhabited for centuries by peoples who had developed techniques that made it possible to survive in the desert. Although the native inhabitants' experience of nature was different from the European perspective that was gradually establishing science as a central component of western civilization, their efforts to survive and prosper in the desert regions required knowledge of the environment and an ability to apply that knowledge as needed.

The Southwest's native population made extensive use of the region's indigenous plants. Although the seeds and roots of several plants, such as the mesquite, served as dietary staples, much more significant was the natives' use of plants for medicinal and domestic products. The Tohono O'odham, for example, employed the leaves of the creosote bush as a treatment for snake, spider, and scorpion bites, chewing the leaves and applying them to the swollen area. Boiled leaves were used as an emetic, while heated branches were wrapped in cloth and applied to painful joints. Mesquite was used even more widely. The Pima boiled the leaves to use as an emetic and prepared a laxative from the boiled bark of the tree. They harvested the gum produced by the mesquite and used it to treat various maladies. A tea made from this product served as treatment for diarrhea and sore throat, while a water solution of the gum was often used to cure indigestion. Sore eyes and open wounds were treated with a dilute solution of boiled gum as well. Although such treatments were rarely systematized, they remained an important aspect of native folkways and were preserved for centuries. During the twentieth century, such treatments remained an important part of native cultures. The Navajo, for example, continued to use numerous native plant varieties for treatment, applying some three dozen externally and ingesting nearly sixty.[1]

Native inhabitants also applied their knowledge of desert plants for other purposes. Throughout the Southwest, mesquite gum was used as mucilage and as a base for a shampoo that darkened hair and killed lice, according to nineteenth-century botanist Edward Palmer. The fibers of several yucca varieties were treated and used to produce ropes, nets, clothing, mattresses, and blankets. The stems of these plants were cut into slices, beaten into a pulp, and mixed with water to serve as soap. Basketry required equally sophisticated knowledge of native plants, as shown by the use of devil's claw by the Tohono O'odham. Although the wild native variety of this plant was used in basketry as early as A.D. 600, the key development was the domestication of this plant in the nineteenth century to take advantage of a variant with a significantly longer claw. The longer and more flexible fibers of this variant were much better for basketry, an industry that became increasingly important as the Anglo population of the Southwest increased. Noting that the longer-clawed plants were marked by white seeds, the tribe began artificially breeding devil's claw by saving these seeds for use in the next growing season. Not only were the native inhabitants of the Southwest able to apply their knowledge of desert plants to specific needs, they were also able to improve certain plants in much the same way as other agriculturalists.[2]

Because of the harsh desert environment, natives of the Southwest were especially concerned with the development of an appropriate agricultural base for their societies. Varieties of corn, beans, and squash were introduced from Mexico well before 1000 B.C., but such foods served primarily as supplements to foods obtained by foraging. During the next few centuries, however, cultivation of these crops increased and resulted in varieties especially well suited to the desert environment. A hybrid corn was developed by Mogollon farmers in New Mexico circa 400 B.C., for example, yielding larger ears with more rows of kernels. Of greater importance, however, this hybrid was more variable and drought resistant, allowing it to grow under varied conditions. It sprouted from deep planting, which prevented germination until late June or early July when summer rains would water the new plants. This hybrid became the base for southwestern agriculture during the prehistoric period, as indicated by its use by both the Hohokam and Anasazi cultures. A similar agricultural innovation was the development of a bush bean to replace the vining varieties and minimize the competition between crops for water.

The region's arid climate determined agricultural techniques from the beginning. Not only were crop varieties developed to be more drought resistant, but cultivation practices were designed to make best use of scarce water. Techniques that would later be associated with dry farming were used throughout the Southwest. Floodplain agriculture was widely employed, as was the planting of crops at the mouths of arroyos. Native peoples developed several conservation practices as well. Check dams, walled terraces, and other arrangements of rocks enabled the Southwest's early agriculturalists to direct runoff, preserve water, and protect crops. Hopi agriculture displayed these characteristics in dramatic fashion, leading an observer in the 1930s to remark that their techniques were identical to those being applied to Navajo agriculture by soil conservation specialists from the federal government. The blue corn that served as the Hopi's primary crop, though difficult to grow, was especially well suited to the short growing season. This variety was kept as pure as possible by careful selection of seed kernels. Hopi women were adept at seed selection, rejecting kernels from ears that appeared to be a mixture of varieties. Their cultivation techniques, based on deep planting of a large number of seeds in soil that received the maximum amount of water, provided a reliable harvest. The combination of appropriate cultivation techniques and suitable plant varieties allowed the region's inhabitants to survive in an inhospitable environment.[3]

Perhaps the most dramatic example of the technical ability of the Southwest's native inhabitants was the development of irrigation techniques. Numerous examples of limited irrigation exist in the region, but the extensive system of the Hohokam in the Salt and Gila River Valleys has long attracted the attention of observers. Although many details of their system remain unclear, by circa A.D. 800 the Hohokam had established a well-developed irrigation network that would support their society for the next five hundred years. The system of canals ultimately stretched for some 150 miles, with the longest single canal in the Salt River area reaching nearly fourteen miles in length. As wide as thirty feet and as deep as ten feet, some canals were lined with clay to minimize leakage. Because the Hohokam did not build reservoirs, their system was designed to direct stream flow or overflow into their fields, a method especially well suited to the Salt and Gila rivers. When they abandoned the region during the 1300s, they left behind an irrigation system that would not be duplicated until the twentieth century.[4]

Native agriculture adapted to the European and later Anglo-American presence in various ways. Crops introduced by the Spaniards were adopted by the region's inhabitants, including wheat, oats, various fruits, chiles, and onions. The Pima, for example, adopted wheat in the early 1700s and within a century this crop rivaled corn as their most important agricultural product. Planting two varieties (Sonora and Australian), the Pima used irrigation to produce a large crop for the increasing market throughout the nineteenth century. By the late 1800s, Pima irrigation systems supported more than 13,000 acres of wheat. As demand increased, plows began to replace the traditional wooden digging implements, allowing a significant expansion of crop yield. Pima wheat thus became an important addition to the tribal economy, as indicated by an 1859 estimate that they sold approximately 250,000 pounds of wheat to the Overland Mail Line.[5]

The centrality of agriculture to native societies was also key to the most dramatic scientific development of the ancient inhabitants of the Southwest, their knowledge of astronomy. The regular movements of the moon, sun, and stars provide observers with an accurate calendar of the changing seasons to guide crop plantings and other activities. The Sun Chief of a Hopi village still watches the sun move north along the eastern horizon, noting its position with respect to natural landmarks. The connection between the sun's position and the likelihood of a damaging frost has been passed down over centuries. Similar horizon observations are also made to determine the dates for festivals and ceremonies during the Hopi religious year. The importance of such astronomical information is indicated by the appearance of the Planet Kachina during the Water Serpent Ceremony in March, a festival that focuses on the planting of corn.

Hopi astronomical observations are important to ceremonies in other ways as well. The November ceremony of Wuwuchim requires careful lunar observations, as the ceremony begins on the first day of the new moon. The timing of rituals within the sixteen-day festival is also determined by astronomy. Making observations through the rooftop opening of a kiva, one of the religious leaders notes the appearance of the first star in Orion's belt, at which point preparations for the specific ritual begin. The ritual itself begins when all three stars in Orion's belt are in alignment with the kiva roof opening; the seven songs within the ritual must be completed before the Pleiades disappear from view. The final part of the ritual is guided by the movement of the stars Castor, Pollux, and Procyon across the kiva roof opening. The knowledge of the appearance and motion of these stars and constellations has been passed from generation to generation as part of Hopi tradition, but it also indicates an understanding of the regularity of celestial motions, even if that understanding has never been systematized into an abstract astronomical model based on mathematics.[6]

The ancient inhabitants of the Southwest, however, appear to have possessed even more sophisticated astronomical knowledge than their modern descendants. The imposing Anasazi ruins in northwestern New Mexico and northeastern Arizona display alignments that indicate an awareness of the summer and winter solstices as well as the vernal and autumnal equinoxes. Observing sites to mark the appearance of these phenomena are complemented by the orientation of numerous structures along astronomical lines. These structures were not observatories as such, but were ritual buildings that reflected the importance of astronomy to the ancient inhabitants of the Southwest. The kiva known as Casa Rinconada in Chaco Canyon, for example, is accurately aligned north and south, but it is also aligned for observations of the solstices and equinoxes. At the time of the summer solstice, sunlight shining through the northeast window at sunrise illuminates specific niches along the northwest wall for several days. Although the kiva is clearly a ceremonial structure, its alignment could only have been possible with a clear understanding of astronomical phenomena.[7]

Elsewhere in Chaco Canyon, the Anasazi constructed an observatory from three stone slabs and two spiral petroglyphs to record solstices, equinoxes, and certain lunar positions. The Fajada Butte structure was designed so that sunlight shining through the spaces between the stone slabs formed characteristic patterns on the spiral petroglyphs to indicate these important astronomical events. This observatory, constructed at some point between 950 and 1150, provided the observational data necessary for the accurate calendar used by the Anasazi for agricultural and ceremonial purposes. Equally impressive, however, the Fajada Butte site also records similar patterns when moonlight shines through the slabs. This alignment would require the Anasazi to know about the 18.6-year cycle of lunar declination extremes, suggesting that they were attempting to reconcile lunar and solar cycles.[8]

The alignment of structures according to astronomical phenomena required careful and continuous observations of celestial phenomena. These observations enabled the Anasazi to construct Casa Rinconada and the Fajada Butte observatory, arranging design elements to record the passage of sunlight through windows and other devices. Their knowledge of astronomy was sufficiently advanced to allow them to construct similarly aligned buildings in locations that did not have visual access to the phenomena of interest. The large ruin of Pueblo Bonito displays several examples of this indirect application of astronomical observations and indicates a sophisticated use of geometry. The north–south alignment of several structural elements is noteworthy, but the east–west orientation of the pueblo's south wall is more significant. This wall coincides with the sunrise/sunset directions on the days of the equinoxes, but the high walls of Chaco Canyon make horizon observations impossible from the Pueblo Bonito site. Thus, the Anasazi architects who designed this structure in the late eleventh century had to use indirect means to orient the pueblo appropriately. Either the orientation was transferred geometrically from some other location or the east-west direction was determined on site by solar observations. Two corner doorways in Pueblo Bonito provide a clear view of the winter solstice sunrise, at which time the sun's rays briefly illuminate the opposite corner of the room. These doors, however, appear to have been interior passages, indicating that the original exterior walls would have blocked the required sunrise view. Once again, the orientation of the doors and the room would require an indirect method of alignment and an impressive knowledge of geometry.[9]

Although Anasazi celestial observations were geared toward ceremonial and agricultural ends, it remains certain that these ancient inhabitants of the Southwest were keenly interested in astronomy. Several pictographs and petroglyphs in the region indicate observations of celestial events, perhaps the most dramatic of which may record the 1054 supernova that formed the Crab Nebula. These artifacts display a crescent moon with a nearby large circular image in an orientation that would have been visible in western North America at the time of the supernova. One of the most noteworthy of these records is located at a sun-watching site near the Penasco Blanco ruin in Chaco Canyon, thus suggesting that the local sun-watchers would have had ample opportunity to observe this rare event. Another possible interpretation of these images, however, is that they record the conjunction of the moon and Venus. Such observations indicate the continued Anasazi interest in astronomical events during their domination of the cultural world of the Southwest from the tenth to the twelfth centuries.[10]

The knowledge and understanding of their environment enabled the native inhabitants of the Southwest to survive and often to prosper in the harsh conditions of the region. The development of appropriate crops and cultivation techniques established agriculture as the foundation for native societies, while the use of native plants for various purposes revealed an understanding of the desert's resources. The continued interest in astronomy, stretching from the ancient Anasazi to the modern Hopi, indicates that observation of nature played an important role in religious ceremonies as well as agricultural practice. Although abstract knowledge of nature was never a goal of the Southwest's native inhabitants, they nonetheless displayed an awareness of their natural surroundings that remained a central part of their culture both before and after European contact.

The Explorers

The Europeans who entered the Southwest in the 1500s brought with them a new perspective concerning the natural world. Although Europe was only slowly entering the period later known as the Scientific Revolution, the explorers who examined the Southwest viewed nature as something to be comprehended, rather than experienced. The most famous of these early explorers was Francisco Vásquez de Coronado. Dispatched north to find rumored wealthy villages, Coronado was also instructed to conduct a general exploration of the Southwest. His large expedition left the Mexico City area in February 1540 and by late summer had surveyed various areas in present Arizona and New Mexico. During the next two years Coronado journeyed as far east as present-day Kansas in an effort to discover the rumored wealth of Quivira and dispatched several exploring parties to survey the Rio Grande valley. The region revealed much of interest. Coronado's first glimpse of elk led him to describe these animals as "sheep as big as horses, with very large horns and little tails." Upon viewing buffalo on the plains, captain of artillery Hernando de Alvarado characterized them as "the most monstrous beasts ever seen or read about." Returning to Mexico City in September of 1542, Coronado's expedition provided the first significant reconnaissance of the region, although it was more of a treasure-hunting effort than a careful scientific analysis. The lack of success finding great wealth retarded further exploration until the end of the century, when such figures as Fray Augustín Rodríguez, Antonio de Espejo, and Juan de Oñate explored much of present-day Arizona and New Mexico. These late-sixteenth-century surveys also added information to the growing knowledge of Spain's far northern frontier, although they can hardly be classified as scientific expeditions.[11]

It was another century before the next wave of exploration penetrated the Southwest. Continuing Spanish/French rivalry spurred Spanish officials to encourage exploration efforts to secure the northern frontier. When Don Diego de Vargas re-established the Spanish presence in New Mexico in the early 1690s, he also participated in a rapid reconnaissance of the region to determine any natural advantages the area might offer. As missionary and other activity increased at the end of the century, knowledge of the frontier again flowed southward to colonial officials, adding to the store of information about the American Southwest.[12]

Perhaps the most noted of the late seventeenth-century explorers was the Jesuit missionary Eusebio Kino. Well trained in science, the Italian-born Kino was particularly adept in mathematics and astronomy. After his arrival in Mexico in 1681, exploration and mapmaking were an important part of Kino's duties as a missionary in Baja California and later in Sonora. His numerous expeditions in the region known as Pimería Alta from 1686 until his death in 1711 provided useful information to establish the Spanish presence and support missionary activity. Kino also observed the natural environment, including the region's flora and fauna, and noted evidence of earlier civilizations. He described the Casa Grande ruin near the Gila River as "a four story building, as large as a castle" and was quite intrigued with the remnants of an irrigation system in the area. Kino increasingly focused on the southwestern portion of Arizona and the northwestern part of Sonora in an effort to determine the practicality of an overland route to Baja California. These surveys also enabled him to establish that California was not an island, as was still widely believed. Descending the Gila River in the fall of 1700, Kino made telescopic observations of the region from a hill near the mouth of the river, from which he could observe the head of the Gulf of California some seventy-five miles to the southwest. Later trips confirmed the peninsular nature of Baja California and led to Kino's 1701 map of the region that clearly showed California as a peninsula. This map was printed in Europe and served as the standard map of the region for several decades.[13]

During the last third of the eighteenth century, a new perspective characterized Spanish explorers in the American Southwest. The Enlightenment focus on the organized pursuit of knowledge provided the Spanish Royal Corps of Engineers (founded in 1711) with a specific function in the Borderlands. In addition to their responsibilities in such practical pursuits as architecture, defense recommendations, dams, and roads, the engineers provided maps and written descriptions of the frontier environment. Although natural history was never a primary function of the corps, observations of flora and fauna found their way into official reports filed in Mexico and Spain. Administrative reorganization in 1776, however, restricted the corps to support functions. The engineers' primary role soon became mapmaking, deemed of greatest importance in Spain's efforts to subdue native groups and potential foreign invaders. Engineering and natural history activities largely disappeared.[14]

The declining fortunes of the Spanish Empire in North America offered opportunities for others to survey the region. During the winter of 1806–7, for example, Zebulon M. Pike of the U.S. Army explored northern New Mexico, later claiming ignorance that he was in Spanish territory. His arrest by Spanish troops in late February resulted in the confiscation of his papers and an escorted trip down the Rio Grande for Pike and his party, before their return to the United States in late spring. Observing the region carefully during this journey, Pike quickly wrote a description of the American Southwest that was published in 1810 as *The Expeditions of Zebulon Montgomery Pike*. A comprehensive discussion of the region, Pike's report stressed the desert nature of the Southwest and emphasized its lack of suitability for agriculture.[15]

The war between the United States and Mexico accomplished more than a political and geographical change in the status of the American Southwest. Eager to gain knowledge of the new territory, the federal government sought to employ a new perspective on the role of science in exploration. The U.S. Army had recognized the important potential role of science in 1838 when it formed the Corps of Topographical Engineers. Usually officered by West Point graduates exposed to a curriculum that included much science, the corps worked closely with leading American scientists such as John Torrey, Asa Gray, and Spencer F. Baird. As a result, the engineers explored the West to establish routes for the nation's expansion, while recording details of the natural history of the region for the American scientific community.[16]

The first opportunity to employ this new view of scientific exploration came at the opening of the war with Mexico. Accompanying Stephen Watts Kearney's march across the Southwest to California during the summer and fall of 1846 were three members of the Corps of Topographical Engineers. Lieutenants William H. Emory, James W. Abert, and William G. Peck were all experienced explorers, but Emory was the key figure in this and later activity in the Southwest. Graduating from West Point in 1831, Emory soon emerged as one of the leading figures in the corps and cultivated relationships with noted American scientists such as Louis Agassiz and Asa Gray of Harvard, Joseph Henry of the Smithsonian, and Alexander Dallas Bache of the U.S. Coast and Geodetic Survey. As a member of Kearney's expedition, Emory combined his primary military function as surveyor and cartographer with the scientific function of a natural historian. He made astronomical and other observations required for accurate maps of the Southwest, but also recorded details of meteorology, geography, and natural history. Emory collected plant specimens in great number, sending them to such leading botanists as John Torrey and Asa Gray for identification and classification. Numerous new plant species were identified from these collections, which were still being analyzed a decade later. Emory's exploration of the Rio Grande Valley and the Gila River route through southern Arizona produced an accurate map of the Southwest and a published report of the region's natural history.[17]

The first large-scale project for the Corps of Topographical Engineers was the Mexican Boundary Survey conducted after the war with Mexico. From the beginning, the survey was plagued by difficulties, most of which stemmed from the clash between corps members and the political appointees who directed the project. Emory (now a brevet major) and his staff were assigned the task of determining the boundary between the two nations, but also hoped to provide collections and observations relating to the natural history of the region. They began their active survey work in late 1850, but when the survey was disbanded two years later numerous uncertainties remained about the boundary. The Gadsden Treaty of late 1853 ultimately solved many of these difficulties and resulted in another commission to survey the boundary. Emory was appointed commissioner, surveyor, and chief astronomer for this commission during the summer of 1854, and spent the next year completing his assignment.[18]

The fieldwork of the two surveys significantly increased knowledge concerning the natural history of the American Southwest. Civilian scientists played an important role in this activity throughout the early 1850s. Botanists Charles C. Parry and Charles Wright made numerous observations and sent specimens to their respective mentors, John Torrey (Princeton) and Asa Gray (Harvard). Zoologist John Henry Clark, a former student of the Smithsonian's Spencer F. Baird, contributed numerous vertebrate specimens, while another Torrey protégé, Arthur Schott, added an extensive collection of insects.

Specimens and observations were sent to eastern scientists who spent several years analyzing, describing, and classifying this material. Noted zoologist Louis Agassiz analyzed numerous fish specimens sent by Emory to the Smithsonian, while geologists such as James Hall used the observations of the surveys to develop a general picture of the geology of the trans-Mississippi West. Dissemination of the results of this work began quickly. Asa Gray arranged for separate publication of Wright's contributions during 1852 and 1853, for example, and performed a similar service in 1854 for George Thurber, another survey botanist. Thurber's collections appeared in the *Memoirs* of the American Academy of Arts and Sciences.[19]

The publication of the official government reports between 1857 and 1859, however, represented the major scientific contribution from the surveys. The first volume (published in 1857) included a general account of the region's geography, climate, geology, flora, and fauna. The geological reports in this volume processed a large amount of data and provided a solid introduction to the geology of the Southwest. Two years later, the second volume of the survey reports appeared, focusing on botany and zoology. Zoological specimens were classified by Spencer F. Baird and Charles F. Girard of the Smithsonian in a descriptive report that included more than 300 different species of mammals, birds, and reptiles. John Torrey's report on the botany of the region was based on his analysis of more than 2,600 species collected by the field parties. A separate report of seventy-five pages by noted botanist George Engelmann discussed the numerous cacti observed during the survey. Together, these reports represented the largest survey of plants in the United States up to that time.[20]

Settlement of the American Southwest soon became a major concern of the federal government. In order to locate suitable routes into and through the region, with special focus on routes to California, the Corps of Topographical Engineers again had an important assignment. When Congress passed the Pacific Railroad Survey bill in early March 1853, it directed the secretary of war to submit a complete report to determine the best route for a transcontinental railroad. This report was to be based on field surveys of the various proposed routes by corps engineers, who were directed to make any observations that might be of use. Such observations were to identify natural resources to support the railroad and the accompanying population that would follow completion of the route. The large number of civilian scientists who were involved with these surveys, either as members of the field parties or as advisors to the War Department, suggested that science would play an important role in the selection of the transcontinental railroad route.[21]

Potential routes included two that traversed the Southwest. The thirty-second parallel route, favored by Secretary of War Jefferson Davis because of its proposed eastern terminus at a point on the lower Mississippi River, would cross the Southwest through southern New Mexico and Arizona. The thirty-fifth parallel route would cross through the northern part of the territory, and was the focus of the first party to take the field. Under the command of Lieutenant Amiel Weeks Whipple, who had served with distinction on the Mexican Boundary Survey, the party included more than a hundred members. The scientific staff included Jules Marcou (a Swiss geologist and protégé of Louis Agassiz), a surgeon-botanist, a physician-naturalist, and several astronomers, surveyors, and engineers. The Whipple expedition reached Albuquerque in October 1853 with instructions to make scientific observations of plant and animal life, the availability of water, the products of the countryside, and the nature of the land itself. Exploring the proposed route during the next four months, Whipple and his colleagues recorded their observations as instructed, noting various plants and animals, some of which were previously unknown. The saguaro cactus particularly impressed the party. Whipple referred to a fourteen-foot high specimen as "immense," while the survey's Prussian-born artist, H. Balduin Möllhausen, denominated the saguaro "the queen of the cactus tribe." Surgeon-botanist John M. Bigelow of Ohio regarded the saguaro as the most interesting cactus of the region and "probably the whole world."

The exploration of the southern route took place in early 1854, led by two parties commanded by Captain John Pope and Lieutenant John G. Parke. The Pope and Parke parties surveyed the eastern and western parts of the route, respectively, and emphasized the level terrain and lack of major obstacles for a railroad line. The shortage of timber and water along the route presented a potential problem but climatic and geographical problems were minimal. Pope and Parke commanded smaller parties and fewer scientists than had Whipple, but these expeditions also provided specimens of plant and animal life as well as descriptions of the region's geology and geography.[22]

The railroad surveys, which included several other routes in the West, brought back a large collection of information. The specimens were added to the existing collections of the Smithsonian Institution, turning that facility into one of the major museums of the world. The time constraints and focus on specific routes limited scientists' observations, a particular problem for geologists, but the published reports that began to appear in 1855 nonetheless provided an overview (however superficial) of the regions surveyed. The seventeen volumes of the *Pacific Railroad Reports* published between 1855 and 1860 provided discussions of flora and fauna, geological descriptions, numerous maps, and hundreds of illustrations. Those expeditions that explored the Southwest contributed greatly to the cumulative knowledge begun a few years earlier by the Mexican Boundary Survey. As a result, knowledge of the natural history of the American Southwest by the late 1850s was arguably more complete than that of other areas of the West.[23]

As the nation drifted toward the Civil War, the Southwest continued to be of interest to those who coordinated military survey activity. Of great significance to science was the Colorado River survey conducted by Lieutenant Joseph Christian Ives of the Corps of Topographical Engineers during the first half of 1858. Ascending the river by steamboat from Fort Yuma to Black Canyon, Ives's party included the well-trained geologist John S. Newberry, who served as surgeon and naturalist for the expedition. A protégé of the famous American geologist James Hall, Newberry was particularly important during the expedition's survey across northern Arizona to Fort Defiance. The first trained geologist to explore the Grand Canyon floor and to see the plateau country of northern Arizona and southern Utah, Newberry made various observations that could be integrated with those made by other geologists elsewhere in North America. Despite the hurried nature of his observations in the Grand Canyon, he traced the stratigraphic column, providing information of growing importance to American geologists. Newberry's observations of the Plateau Province were equally significant. He concluded that the valleys and mesas of the region were a result of "a vast system of erosion, and are wholly due to the action of water." Newberry's contributions to geological knowledge were based on his superior training and his ability to generalize from observations. He was thus able to integrate information from the field with more theoretical concepts being considered by his colleagues in the leading institutions of the East. Newberry's efforts marked an appropriate end to the surveys of the antebellum American West, as they pointed toward a more sophisticated scientific outlook.[24]

Following the Civil War, the American West again became a focus of exploration involving scientists from a number of agencies of the federal government. The Smithsonian Institution, Coast and Geodetic Survey, and the Departments of War, Agriculture, and the Interior all viewed the West as a source of new data, but also hoped to integrate these data into a broader scientific view of the region. Much of this work involved surveys conducted by army personnel, continuing the pre-war efforts of the Corps of Topographical Engineers (disbanded during the Civil War). Civilian scientists nonetheless played crucial roles, either as members of such army expeditions or as independent explorers.[25]

Probably the most famous southwestern explorer of the post–Civil War period was John Wesley Powell. His dramatic journey through the Grand Canyon in 1869 attracted much attention, but as a scientific expedition this journey was very limited. The Smithsonian Institution encouraged him in his work and he was authorized to draw government rations for the trip, but Powell apparently made no effort to recruit trained naturalists to accompany him. As a result, the scientific contributions from this trip were restricted to Powell's limited geological observations and his comments concerning the role of erosion in shaping the region. This journey did, however, convince Powell that a more extensive and scientific survey of the region was required. He persuaded Congress to appropriate $10,000 for a "Geographical and Topographical Survey of the Colorado River of the West," but again failed to include trained naturalists in the party that explored the region between May 1871 and February 1873. His own observations were largely limited to topography and ethnology, although the expedition leader was only intermittently attached to the field party during this period. Powell's later U.S. Geographical and Geological Survey of the Rocky Mountain Region (usually referred to as the Powell Survey) explored the Plateau Province with the same limited perspective.

Powell's observations of the arid Southwest, however, displayed an understanding of the unique characteristics of the region. Describing the desert of southwestern Arizona in 1895, he noted that the "landscape of vegetal life is weird—no forests, no meadows, no green hills, no foliage, but clublike stems of plants armed with stilettos." A quarter century after his dramatic journey down the Colorado River, the Grand Canyon region continued to fascinate the explorer. "All the scenic features of this canyon land are on a giant scale," he wrote, "strange and weird." Powell's most important scientific contribution also stemmed from his recognition of the region's uniqueness. His interpretation of geological observations within the context of uniformitarian concepts led him to stress the importance of erosion as the key feature of the province.[26]

The scientific exploration of the Southwest was better served by the efforts of George M. Wheeler during the 1870s. A graduate of West Point in 1866, Wheeler coordinated an extensive survey of eastern Nevada and extreme northwestern Arizona during the summer and early fall of 1871. Although this survey was designed to focus on practical concerns such as transportation, settlement, and military needs, science was not neglected. On Wheeler's staff was the geologist Grove Karl Gilbert, who made numerous observations along the upper Colorado River and in the area north of the Grand Canyon. These observations enabled Gilbert to note the volcanic nature of the San Francisco Peaks and to emphasize the role played by water erosion in forming the topography of the region. Gilbert would soon emerge as one of the nation's leading scientists and served as chief geologist for the U.S. Geological Survey in the early 1890s.

In 1873, Wheeler was placed in charge of the last great military reconnaissance, the Geographical Surveys of the Territories of the United States West of the 100th Meridian. The Wheeler Surveys, as they were more commonly called, examined much of the American West and included the examination of most of Arizona and New Mexico. Scientists working on these surveys at one time or another during the 1870s included Gilbert, paleontologist Edward Drinker Cope, and zoologist Elliott Coues, all of whom made significant contributions and became well known figures in the development of American science. Although the focus of the Wheeler Survey was accurate mapping of the region and the evaluation of settlement possibilities, science played an important auxiliary role. When the U.S. Geological Survey was formed in 1879 to consolidate and coordinate western exploration, science remained an adjunct to the survey's goal of an accurate catalog of natural resources. As a government function, exploration was expected to have a practical benefit. Scientists serving with the field parties often made interesting scientific observations, but these efforts were subordinate to the knowledge necessary to promote settlement and development in the American West.[27]

The Naturalists

Although official government surveys viewed science as an adjunct to more practical concerns, scientists in the late nineteenth century enjoyed numerous opportunities in the American Southwest. Geologists, zoologists, botanists, and other scientists journeyed throughout the region, making observations and recording their conclusions in a variety of scientific publications. Some of these naturalists were associated with government expeditions, others represented eastern museums and agencies, while a few worked as independent collectors. The material generated by these observers provided a more complete picture of the region's natural history and offered additional insight concerning various scientific topics.

Among those who contributed to the growth of knowledge of the Southwest were army surgeons who often had great interest in natural history. One of the most famous of these surgeon-naturalists was Elliott Coues, a protégé of the Smithsonian's Spencer Baird. Following completion of medical training in 1863, Coues was commissioned as assistant surgeon with the army and posted to the Department of New Mexico in the spring of 1864. He probably owed this assignment to his mentor Baird, who recognized that the Southwest remained a region of great opportunity for naturalists. Coues arrived in Santa Fe in June and was assigned to Fort Whipple, Arizona Territory, by department commander General James H. Carleton. The young naturalist enthusiastically pursued his interests while serving as post surgeon until he left Arizona in late October, 1865.[28]

During his service in Arizona, Coues observed and collected numerous animal specimens, but his work as an ornithologist established his reputation. By the end of his Arizona assignment, he had collected some 600 specimens and had already published two articles on his work in the British ornithological journal the *Ibis*. He spent the first half of 1866 working with Baird at the Smithsonian and writing articles detailing his observations of southwestern birds. He prepared an extensive treatment for the Academy of Natural Sciences of Philadelphia, who published his "List of Birds of Fort Whipple, Arizona" in their March 1866 *Proceedings*. Far more than a mere catalog, Coues's discussion also described the characteristics of Arizona birds and compared these characteristics with those of specimens elsewhere in the Southwest. The climax of his discussion of Arizona bird life was *Birds of the Colorado Valley*, published in 1878 by the U.S. Geological Survey of the Territories. This volume provided careful descriptions of the region's avifauna and was based on Coues's own collections as well as those in the Smithsonian.[29]

Coues also contributed to knowledge of other animals found in Arizona.

His collection of forty-four species of reptiles and amphibians, for example, provided the base for an 1866 article published by Edward Drinker Cope in the *Proceedings* of the Academy of Natural Sciences of Philadelphia. During late 1867, Coues published an extensive discussion titled "The Quadrupeds of Arizona" in four issues of *American Naturalist*. Stressing the opportunities for natural history observations in the "wild and primitive region which constitutes the Territory of Arizona," Coues introduced his study by observing, "The traveller meets, at each successive day's journey, new and strange objects, which must interest him, if only through the wonder and astonishment they excite." His discussion of the mammals of the territory, ranging from rats and mice to mule deer and antelope, covered forty pages and was augmented by a brief discussion in the *Proceedings* of the Academy of Natural Sciences of Philadelphia.[30]

The opportunities for army surgeons interested in natural history were also displayed by the service of Edgar A. Mearns in the Southwest during the 1880s. A graduate of the College of Physicians and Surgeons in New York, Mearns spent several months as temporary curator of ornithology at the American Museum of Natural History before gaining a commission as an army surgeon in late 1883. Free to choose from a list of posts, Mearns decided to accept assignment to Fort Verde, Arizona Territory, because of its greater opportunities for natural history. He spent nearly five years in Arizona, making numerous observations and collections, most of which focused on the mammals and birds of the territory. Mearns accompanied cavalry detachments on their missions throughout the Southwest, which enabled him to survey the natural history of much of Arizona, New Mexico, and extreme west Texas.[31]

Mearns's observations of the animals of the region resulted in a large number of publications in the scientific press. From one of his earliest extensive trips in the region, a spring 1885 journey from Fort Verde to Deming, New Mexico, he contributed an account of the mammals of the region to the *Bulletin* of the American Museum of Natural History. His essay focused on a relatively unknown species of squirrel, only a few specimens of which had been known before Mearns's work. Not only did his discussion provide more details of the characteristics and habits of this species, it also established that the habitat of this squirrel was much larger than originally thought. Discussions of otters, skunks, weasels, and foxes appeared later in the same publication, as did an extensive article on new species and subspecies of mammals identified by Mearns during his years in Arizona. In these publications, Mearns was careful to integrate his own observations with those of other naturalists who had examined the region in earlier years.[32] Similarly extensive discussions of the bird life in the region appeared in the noted ornithological journal the *Auk* and were published soon after his return from Arizona.[33]

Mearns spent late 1889 and early 1890 at the American Museum of Natural History, where he worked on the manuscripts that would be published during the next two years. Another opportunity to explore the American Southwest arose in late 1891 when he was appointed medical officer of the Mexican–United States International Boundary Commission. Through a cooperative arrangement between the commission, the Smithsonian Institution, and the American Museum of Natural History, Mearns explored the entire boundary line from El Paso to the Pacific coast before the work of the commission ended in September 1894. Mearns oversaw the collection of some 30,000 specimens, most of which were sent to the U.S. National Museum in Washington, D.C. Following his return from the Southwest, he began to organize the material he had sent from the field and to write various accounts of his observations. The most significant of these publications appeared many years later as *The Mammals of the Mexican Boundary of the United States*, published as a bulletin of the U.S. National Museum in 1907, slightly more than a year before his retirement.[34]

The American Southwest also provided opportunities to naturalists who worked as individual collectors. Among the most intriguing of these collectors was the husband and wife team of John Gill and Sara Plummer Lemmon. A native of Michigan, John Lemmon came to California in 1866 to recover his health following imprisonment in Andersonville during the Civil War. By the late 1870s, he was making a suitable living running a small private school, selling plant specimens to eastern collectors and museums, and writing articles on California plants for local newspapers. During his botanical expeditions, Lemmon met Sara Allen Plummer, who shared his interest in botany and displayed great talent as a plant portraitist. After exploring together in California for several years, the couple married in 1880 and took up residence in Oakland, where they remained active in civic and scholarly activities when they were not traversing the West in their quest for new botanical specimens.[35]

The Lemmons made several trips to Arizona, beginning with a preliminary excursion to the Santa Catalina Mountains and the Tucson vicinity in the spring of 1880. Following their wedding in November, the Lemmons decided to return to Tucson for their honeymoon trip so they could explore the Catalinas in greater detail. Arriving in Tucson in early March 1881, they quickly established a foothills base camp in an abandoned cabin and located a cave at a higher elevation to use as an advance camp. Despite the discomfort, they frequently spent the night in the cave to allow greater time for collecting and observing during their two-week expedition. From either camp, however, the Lemmons faced difficult terrain in their search for specimens. As John wrote from Tucson, they arduously climbed "the rugged steeps, contending all the way against the thorns of the mesquite, the bayonets of yucca and the fiendish needles of cacti, each of these terrible defensive plants being found clinging to the very pinnacles of these mountains, as if guarding vast treasures."

Although their expedition provided them with access to numerous rare and new plants, which they dried for later study, their efforts to penetrate deeper into the Catalinas proved unsuccessful. In mid-March, they abandoned their camps and returned to Tucson, where they learned that the unclimbed mountain peak that had been their goal could more easily be reached from Oracle, on the north side of the mountains. They spent the next two weeks preparing for their ascent and making botanical collections in the area. Over a three-day excursion, the Lemmons recorded observations of plants and animals and finally reached the summit of the peak that would later be named in their honor. During the next month they explored other areas of southern Arizona, visiting the Huachuca and Chiricahua Mountains, Fort Bowie, Apache Pass, and several other sites before returning to California in early May. John Lemmon spent the summer writing newspaper and magazine articles about this Arizona trip and presented a paper on "Four New Trees of Arizona" to the California Academy of Science.[36]

The Lemmons returned to Arizona in each of the following three years. They explored the area surrounding Fort Huachuca during the summer of 1882, using the fort as their headquarters for protection against Apache raids in the region. Upon their return to California, they gave a series of talks on their latest adventures, including a program that attracted nearly 700 people to hear the Lemmons discuss the "Perils and Pleasures of

Botanizing in Arizona." They further explored the Catalinas during the 1883 field season, but spent the next year's visit investigating the flora and fauna of northern Arizona. The Lemmons observed and collected over a wide variety of habitats, including Mohave County, the Grand Canyon, Oak Creek Canyon, the San Francisco Peaks, and the region surrounding Flagstaff. During their five Arizona trips the Lemmons discovered many plants previously unknown, which were cataloged and classified by such eastern botanists as Asa Gray. They thus contributed to a better understanding of the region's flora and saw their work well received by scientists elsewhere in the nation.[37]

The birds of southern Arizona also offered a focus for independent naturalists, as shown by the activities of Tucsonan Herbert Brown. Born in Virginia, Brown moved to Tucson in 1873, where he pursued mineral prospecting and various positions with the city's newspapers. He later served as Clerk of the Superior Court of Pima County. Intrigued with the natural history of the region, Brown became an active bird collector in the early 1880s and was elected an associate member of the American Ornithologists' Union in 1885, gaining full membership in 1901. He served as curator of the Arizona State Museum at the University of Arizona beginning in the early 1890s, donating his collections to establish the facility.

Collaborating with other naturalists who visited the Tucson area, Brown soon learned how to preserve and identify birds. His interest quickly led to publications describing little-known species and attracted much interest from the ornithological community. His account of the masked or Arizona bobwhite in *Forest and Stream*, for example, was cited in an 1886 article written by noted naturalist J. A. Allen for the *Bulletin* of the American Museum of Natural History. Allen stressed that this species had only been known since 1884 and that all but one of the nineteen specimens currently cataloged had been collected by Brown along the Arizona/Sonora border. Brown's own publications provided discussions of numerous species throughout the late 1800s and early 1900s. His articles in the *Auk* included detailed descriptions of several species of thrashers, owls, and buzzards, as well as a retrospective essay on the masked bobwhite written in 1904. When he died in 1913, Brown was serving as president of the Audubon Society of Arizona and was enjoying a respectable reputation among his fellow ornithologists.[38]

Although the American Southwest offered the individual collector great opportunities, the region also attracted the attention of institutions that were intrigued with potential new knowledge. These institutions encouraged their personnel to travel to the Southwest, either officially or unofficially, to survey the natural history of the region. Joseph F. James, custodian of the Cincinnati Society of Natural History, spent six weeks in the Tucson area during the spring of 1881, examining the unique desert vegetation in the area. The results of his survey were published at the end of the year in *American Naturalist* under the title "Botanical Notes from Tucson." James described numerous plants of the area, paying special attention to cactus species such as the ocotillo, which he described as "certainly one of the most striking of all found on the deserts of Arizona." Other observations and collections included desert shrubs and trees as well as other cacti, many of which were described and illustrated in the article.[39]

During the late summer of 1889, the U.S. Department of Agriculture coordinated a biological survey of the San Francisco Peaks in northern Arizona. Led by Clinton Hart Merriam, chief of the department's Division of Ornithology and Mammology, the survey team included such noted naturalists as Merriam's colleague Vernon Bailey, paleontologist F. H. Knowlton of the U.S. Geological Survey, and Leonhard Stejneger, curator of reptiles at the U.S. National Museum. During August and September, the survey recorded the distribution of species in the region and compiled annotated lists of the mammals, birds, amphibians, and reptiles found there. The official report published in September of 1890 proved especially valuable for its division of the region into seven life zones, each of which was discussed in detail. Descriptions of the characteristic flora and fauna in each zone were accompanied by comments concerning the geographical characteristics of northern Arizona, providing a comprehensive overview of the region's natural history and environment.[40]

Local institutions often contributed to the advance of knowledge concerning the Southwest as well. Botanist J. W. Toumey of the recently established University of Arizona, for example, spent July and August of 1891 collecting and studying the plants of central Arizona. Describing his activity for the *Botanical Gazette*, Toumey focused much of his attention on the small mining camp of Big Bug, some eighty miles north of Phoenix. During his two-day visit, he was struck with the wide variety of vegetation in the area and collected seventy-five plant species, all in fit condition for herbarium specimens. Flowering plants and conifers were especially noteworthy in this collection, leading him to describe the area as containing "as interesting a bit of flora as I have seen since coming to the territory." Entomologist T.D.A. Cockerell of the New Mexico Experiment Station made extensive observations of bees and flowers during the mid-1890s, discovering a unique aspect of this relationship in New Mexico. He informed readers of the *Botanical Gazette* that in contrast to the situation in Europe and eastern North America, it was quite common in New Mexico to find species of bees "practically confined to particular species of flowers." The practical concerns of the experiment station could lead to discoveries of interest to naturalists as well as agriculturalists.[41]

The American Museum of Natural History conducted its own expeditions to the Southwest in the late nineteenth century. During the first nine months of 1894, a small party led by the museum's W. W. Price engaged in an extensive collection of mammals that ultimately included some 1,500 specimens. Most of the specimens were collected in Cochise County, although the party also collected in Pima, Graham, and Apache counties, as well as in northern Sonora. Their specimens came from a variety of terrains and included numerous examples of the region's rodents, in addition to larger mammals such as deer, mountain lions, and foxes. The specimens were sent back to the New York museum for identification and classification, although half of the collection was later sold to the Field Museum in Chicago.[42]

Early in the following century, the U.S. Bureau of Fisheries dispatched well-known zoologist Frederic Morton Chamberlain to conduct a survey of the fishes of Arizona Territory. Between mid-January and late April 1904, Chamberlain concentrated his attention on the Gila River basin, but also observed the fish population of the Santa Cruz and San Pedro rivers and the Colorado River near Yuma. Far more than a catalog of the territory's fish, his report noted the changes in the fish population over the previous few decades, comparing previous observations with his own. Chamberlain was struck by the disappearance of fish in many locales, likely the result of the visible environmental decline. He suggested that overgrazing and other unwise agricultural practices, overcutting forests and woodlands, pumping groundwater, and various mining activities combined to threaten the natural habitat. Chamberlain was quick to point out that blame for the disappearance of the region's fish could not be placed solely on one cause, but the cumulative effect of the stresses on the environment were proving fatal. The solution to this problem lay in pond culture, as current climatic and economic conditions made restocking streams impractical. "With a more general knowledge of pond culture," Chamberlain concluded his report, "the necessary natural depletion of Arizona streams may be viewed with equanimity."[43]

The Southwest in the early twentieth century was also the focus of significant interest from the U.S. Biological Survey of the Department of Agriculture. Originally established in 1885 as a unit within the entomological division, the survey was designed to provide information to farmers concerning native animal species. By the time of its separate creation as the Bureau of Biological Survey the unit had become noted for its comprehensive and systematic collection of birds and mammals. Directed by Clinton Hart Merriam from its beginning, the survey was plagued by inadequate funding and was forced to rely on consultants and independent collectors for much of its field activity. As indicated by the 1889 survey of northern Arizona, however, the unit also had a small staff of scientists such as Vernon Bailey who, by the late 1890s, had become the survey's chief field naturalist.

Shortly after accepting this position, Bailey married Florence Merriam, younger sister of the survey director. Merriam had attended Smith College as a non-degree student in the early 1880s, pursuing her interest in natural history. During the mid-1890s, she made several trips to the West in an effort to regain her health (Merriam probably suffered from a mild case of tuberculosis), making observations of the region's birds on every occasion. Her observations in Utah, Arizona, and especially California resulted in several published accounts that established her as a respected nature writer with a special interest in ornithology. Her marriage to Bailey created a gifted team of naturalists. Bailey's chief interests in mammals, reptiles, and plants were augmented by his wife's interest in birds. For more than thirty years after their marriage in 1899, the Baileys remained active field naturalists.[44]

The Baileys' investigation of the Southwest involved annual visits to New Mexico during the spring and summer field seasons of 1900–1906. The 1900 and 1901 surveys concentrated on the Big Bend region of Texas and the Carlsbad Caverns area of southeastern New Mexico. Florence's observations of the region's birds were added to those she had already made elsewhere in the West and served as the foundation for her first major publication, *Handbook of Birds of the Western United States*, a field guide published in 1902 by Houghton Mifflin Company. When Vernon Bailey was assigned to make a detailed biological survey of New Mexico Territory, Florence viewed the prospect as an opportunity to expand her southwestern observations, update the *Handbook*, and write magazine articles for a wider audience.

During the next three summers, the Baileys explored virtually the entire territory, identifying and collecting numerous species of flora and fauna. In an extensive discussion of their 1903 fieldwork, Florence Bailey provided readers of the *Auk* with a description of the upper Pecos region and provided details of ninety-three species of birds observed in the area. Similar observations were made elsewhere in the territory, ending in the fall of 1906 at a mountain camp near Mogollon in the extreme western part of the territory. The high altitude of this camp (8500 feet) provided the Baileys with significant new information concerning the distribution of the territory's fauna and yielded numerous specimens of birds and small mammals that were shipped to the Biological Survey headquarters.[45]

The Baileys continued as active members of the Biological Survey well into the 1930s and often contributed additional knowledge concerning the American Southwest. During the spring of 1921, for example, they surveyed the Santa Rita Mountains in southern Arizona, paying special attention to migratory birds that were returning to the region and those that were migrating through the mountains. Their camp provided the Baileys with ample opportunity for observations. So many birds passed through the area, Florence later wrote, "that the first thing in the morning I looked from the tent door to the sycamore tops, after which I went down to the ranch to look in the live oak tops to see what had come in the night." Her observations were not limited to birds, however, as the landscape continued to fascinate her. As early as mid-March, "a desert tree was delighting our eyes on our trips down through the giant cactus country— the palo verde, a solid mass of vivid lemon-yellow bloom, gorgeous in the sunshine." Indeed, the spring vegetation was the focus of the concluding paragraph of Florence's published account of the Santa Rita survey. "The mesquites which had now come into fresh green leaf," she rhapsodized, "were fragrant with yellow tassels. And here and there big cactus flowers were to be seen. The desert was putting on bridal garments."[46]

Florence continued to publish articles describing the Baileys' work, further establishing herself as one of the leading ornithological writers of her time. Since 1916, she had been working on an extensive discussion of her observations in New Mexico, seeking to provide as complete a description as possible of the state's bird population. Published in 1928 by the New Mexico Department of Game and Fish, Bailey's 800-page *Birds of New Mexico* was well received by the ornithological community. By the time of her death in 1948, Florence Merriam Bailey had become one of the best-known ornithologists of the twentieth century and had contributed significantly to a better understanding of the natural history of the American Southwest.[47]

The work of naturalists in the nineteenth and twentieth centuries emphasized the status of the American Southwest as a source of new information. Long an important part of the development of science, the collection of new species of plants and animals in the Southwest took on special significance because of the region's unique environmental characteristics. The arid climate impressed observers both as a regional characteristic and as a possible explanation of the uniqueness of the region's life forms. These naturalists, however, were in many respects carrying on the tradition of earlier explorers who were no less struck by the region's natural characteristics. Although the early Spanish explorers and missionaries were less systematic in their work than later naturalists, their observations were no less valuable in alerting the scientific community of the time to the existence of new and unusual plants and animals, as well as to a region with unique physical characteristics. The Southwest's existence as a natural laboratory established the region as a scientific resource that offered numerous opportunities for the advance of knowledge.

Chapter 2 Applying Science

Benjamin Silliman Jr., Mining Consultant

Because of the region's abundant natural resources, the extractive industries played a crucial role in the economic development of the American Southwest. Mining, ranching, forestry, and commercial agriculture all contributed to this development and often defined the region for observers in other areas of the nation. Equally important, science proved central to each of these activities. Plant and animal breeding led to significant changes in farming and ranching, while the scientific management of natural resources led to more efficient timber harvests and water distribution.

Especially during the late nineteenth century, scientists played a major role in the growth of the Southwest's mining activity through their service as consultants. Traveling throughout the region, these scientists advised mining companies on potential mineral deposits and suggested new techniques for the recovery and processing of ores. Such advice was, to be certain, based on the analysis of samples and comparisons with outwardly similar deposits elsewhere; estimates were thus all these consultants could provide. These estimates, however, could play a significant role in the financial aspects of mining endeavors. Geologists and chemists surveyed properties and prepared reports for their employers, who paid generously for such scientific expertise. If favorable, these reports often found their way into the company's prospectus to encourage prospective investors. Positive comments from scientific experts frequently made the difference between a mining company's initial success and failure.[1]

Among the most active of these scientific consultants was Benjamin Silliman Jr. (1816–1885). The son of the noted Yale chemist Benjamin Silliman (1779–1864), the young man showed an early interest in science and was part of the American scientific community almost as a birthright. The elder Silliman had played a crucial role in the foundation of an American scientific community during the first half of the nineteenth century, primarily through his efforts as editor of the *American Journal of Science and Arts* (popularly known as "Silliman's Journal"). This periodical served as a means of communication among the nation's scientists and augmented the institutional growth that led to national organizations. Although Silliman's scholarly contributions were modest, his efforts to build a scientific community gained him a national reputation and the recognition of his colleagues.[2]

Following a traditional education in the schools of New Haven, the younger Silliman attended Yale. He received his undergraduate degree in 1837, but weaknesses in the science curriculum required Silliman to study independently with his father and other Yale mentors. He continued this self-directed scientific education after graduation, serving as an assistant in his father's laboratory and classroom, pursuing an A.M. degree at Yale, and briefly studying in the Boston laboratory of Charles T. Jackson. Silliman's study in Boston introduced him to the methods used by mineral assayers such as Jackson, who was also quite active as a scientific consultant in New England. Upon completion of his graduate degree in 1840, Silliman served in various capacities at Yale, gaining appointment as "Professor of Chemistry and the Kindred Sciences as Applied to the Arts" in 1846.

To augment his meager salary, Silliman soon became involved in consulting activity. His scientific reputation grew throughout the 1840s, leading to additional consulting opportunities and appointment to the faculty of the University of Louisville Medical Department in 1849. He taught in Kentucky until the spring of 1854, when he assumed his father's position as chemistry professor in the Yale medical faculty. Despite the additional income from these academic posts, Silliman remained an active

During the 1860s and 1880s, Yale chemist Benjamin Silliman Jr. visited the Southwest on several occasions in his role as a mining consultant. (Courtesy of the National Museum of American History, Smithsonian Institution)

consultant, examining coal mines in New Brunswick and Kentucky and offering his expertise to various businesses in New England. By nature gregarious, optimistic, and trusting, he remained a popular figure among those seeking professional expertise. These same characteristics, however, would ultimately involve Silliman with several questionable ventures which damaged his reputation.[3]

Silliman's consulting activity paralleled that of many American scientists in the 1850s. The practical knowledge offered by these consultants was valued and rewarded, attracting such leading American scientists as Josiah Dwight Whitney, James Hall, J. Peter Lesley, and others. In contrast to faculty salaries (generally less than $2,000 per year in the mid-1850s), consulting activity paid handsomely. Whitney commanded fees of $500 per week during the 1850s, while Lesley turned down faculty chairs at Columbia and Cornell because he could make significantly more money as a consultant. Silliman's efforts were similarly remunerative, but he also benefited professionally by publishing the results of his investigations in *American Journal of Science* and other periodicals.[4]

The most significant of Silliman's consulting activity during the 1850s involved his analysis of petroleum deposits in western Pennsylvania. The oil seepages in Venango County had attracted attention early in the decade, leading to the formation of the Pennsylvania Rock Oil Company in September of 1854. Among the investors in this company were several individuals from New Haven, who wanted a scientific analysis of the petroleum. These investors knew of Silliman's reputation and convinced the company to hire the Yale chemist. Silliman used fractional distillation techniques to separate the constituent parts of the sample, completing his formal report the following April. Silliman suggested that the Venango County petroleum could be used as an illuminant, a relatively novel idea at a time when petroleum served chiefly as a lubricant. Silliman's report, for which he was paid slightly more than $500, encouraged the company's investors. They expanded their operations and later hired Edwin Drake to begin the drilling efforts that led to the nation's first successful oil well four years later.[5]

Among the individuals Silliman met during his petroleum work was Thomas A. Scott of the Pennsylvania Railroad. During the next few years, Scott became famous as a transportation specialist, serving the government in various capacities during the Civil War and continuing to oversee the Pennsylvania Railroad and his own investments. These investments included several mineral claims in the extreme northwestern part of Arizona Territory. Scott approached Silliman in early 1864 to evaluate these and other western properties. The chemist would be free to accept assignments from others as such opportunities presented themselves, suggesting to Silliman that this trip could be a very profitable one.[6]

At the time of Silliman's western journey, Arizona was already known for its potential mineral wealth. Prospectors had discovered gold deposits along the Colorado River and in the Prescott area, precipitating a significant population influx and creation of Arizona Territory in 1863. Mineral deposits in the Fort Mohave region of northwestern Arizona were brought to the attention of San Francisco capitalists by noted prospector Johnny Moss (referred to by some as the "Kit Carson of Mining"). Moss's discoveries led to the incorporation of dozens of mining companies, most of which were speculative ventures, and the formation of the San Francisco Mining District during the spring of 1863. Among those with claims in the district were prominent San Francisco mining promoters Levi Parsons and George Hearst.

Parsons, who had arrived in California in 1849 to practice law, had made a small fortune as a speculator in mines and land. He telegraphed news of the Moss discoveries to friends in the East, convincing Thomas A. Scott to organize the Arizona Gold & Silver Mining Company to capitalize on the strikes. Scott was also involved with the Philadelphia Ophir Company and the Philadelphia Silver & Copper Mining Company, the former of which purchased the Moss Lode in late 1863. He soon organized and sent west a twenty-one-man expedition to examine the Fort Mohave area. This party was led by John Wyeth (later one of the founders of Wyeth Pharmaceuticals) and George Noble, but arguably the key figure in this group of "enterprising Pennsylvanians" (as described by a writer for the *Arizona Miner*) was Thomas Scott Stewart. Nephew of Thomas A. Scott, Stewart was elected recorder of the San Francisco Mining District shortly after the party reached the area in early April 1864. Stewart, Wyeth, Noble, and Parsons would all play important roles in Silliman's western expedition.[7]

Before leaving New Haven in mid-March of 1864, Silliman arranged to examine properties in the West for several eastern investors. He arrived in San Francisco on 9 April and traveled extensively in California and Nevada during the next three months. Signing additional contracts with local mining promoters, Silliman examined mines in Virginia City, Nevada, and such California locations as Placerville and the Mother Lode region of Sierra and Nevada counties. He visited the New Almaden quicksilver mine south of San Jose, another property of Thomas Scott, and also surveyed such sites as Bodie, Aurora, and Reese River, all in Nevada. Making use of his earlier experience in the petroleum business, Silliman signed a $2,500 contract with E. Conway & Company to survey oil lands and leases east of San Buenaventura. He completed this task during June, issuing a very favorable report to the company.[8]

Silliman was now ready to begin the survey of Scott's claims in the Fort Mohave area. He left Los Angeles on 15 July, traveling in an army ambulance with Parsons (who served as Scott's legal counsel in the West), Wyeth, and Noble, accompanied by a military escort. The party traveled through Cajon Pass and across the Mohave Desert to the fort, providing Silliman with ample opportunity to observe the landscape of the region. Struck by the stark character of the desert, Silliman recorded his impressions in an account that was later published. Referring to the Mohave River as "a river in name only" he focused on the lack of water on this route across the desert. He thus described in detail Soda Spring, some ninety miles west of Fort Mohave, as a mineral-rich spring "bearing, among the ignorant guides of these desert regions, the reputation of containing arsenic or some other deadly material." Such was clearly not the case, Silliman wrote, as he and his party drank from the spring "with no ill effects to man or beast, and were very glad to obtain so potable a water after several days of great dearth of this essential of comfort."[9]

Silliman's detailed observations of the mineral deposits of the region included his analysis of the potential for successful mining. Focusing his attention on the San Francisco District, Silliman described several lodes of interest to his eastern clients. The Moss, Skinner, and Parsons lodes were very promising quartz deposits, but several other lodes in the district were noteworthy. Although only preliminary work had been done, Silliman reported that free gold in the quartz was recoverable and in the case of the Moss Lode, assay of selected ore samples yielded more than $4,000 per ton in precious metals. He also stressed the surface similarities between the Moss Lode and that portion of Nevada's Comstock Lode at Gold Hill, suggesting that, as in Nevada, a rich silver vein might be found. During his ten-day visit, Silliman examined some fifty lodes or veins in the

San Francisco District, noting their great potential for profitable mining activity. He made more cursory examinations of other mining districts in northwestern Arizona, but found little of interest. Accompanied by Wyeth and Parsons, Silliman left Fort Mohave during the second week in August. His visit to the region had attracted much interest among speculators and promoters, but mineral production in northwestern Arizona failed to realize the potential Silliman optimistically described.[10]

During their Fort Mohave journey, Silliman and his traveling companions discussed his recent survey of the oil lands of southern California. In contrast to the Arizona mineral lands, the oil properties seemed to offer more immediate promise. When they returned to California in August, Wyeth, Parsons, and other Scott representatives shifted their attention from gold and silver to petroleum. Over the next seven months they purchased a quarter of a million acres of potential oil lands, primarily in the Ojai area of Ventura County, and secured an additional 200,000 acres in leases. Scott and his associates formed a large number of companies to exploit these and other potential deposits, precipitating a frenzy of speculative activity in southern California. Indeed, this southern California oil boom may well have been the most dramatic result of Silliman's western visit. His favorable reports, often published in the promotional tracts issued by these companies, seemed to suggest large returns on investment.[11]

Although Silliman had planned to return east through Idaho and Colorado, the death of his father in December canceled these plans. Silliman left California in early January of 1865, following what he undoubtedly considered a very successful trip. Soon after his return to New Haven, however, Silliman's western adventures took on a more negative tone. The continued speculative frenzy in California oil properties, despite the lack of any significant production, attracted the attention of many observers on the West Coast. In late January the *Mining and Scientific Press*, published in San Francisco, discussed the interest shown by Eastern investors in an article headlined "Swindling Mining Schemes at the East." Although various mining schemes were discussed, the article focused on the petroleum boom and Silliman's role in it. Relying on rumors from unnamed sources, the author wrote that Silliman had realized nearly $200,000 in gold during his trip west and often regulated his price according to how favorable his report was. The following week, the editor of the *Press* retracted his comments concerning Silliman, acknowledging that his fees were much less than reported. Equally important, the editor noted that some of the oil properties in the Santa Barbara area favorably described by Silliman did in fact have potential for profitable development.[12]

Silliman's problems failed to disappear despite this retraction. The speculative bubble in California oil properties burst as quickly as it had begun when Scott, responding to a drop in gold prices, stopped his oil purchases and precipitated a collapse. Many blamed Silliman for this crash, arguing that his excessively optimistic reports had created false hopes. Many of his favorable judgments on other properties had also failed to produce anything other than promotional schemes. Yale faculty were increasingly troubled by his involvement with such speculative activity and some even suggested that he had compromised his scientific integrity for financial gain. Silliman's frequent lack of caution in interpreting analyses and observations had placed him in a professionally awkward position from which he never fully recovered.[13]

Yet Silliman remained an active participant in the efforts to bring eastern financial and scientific expertise to bear on the development of western mining. He gave a lecture in New York in early March 1865, discussing his recent trip and emphasizing the need for more efficient mining methods to attract capital. Shortly after the *New York Times* published a report of Silliman's address, the Yale chemist examined the problem of excessive speculative activity in a letter to the editor. Focusing on mineral activity in California and Nevada, Silliman criticized those corporations that had only been established for speculative purposes, but again stressed the need for outside capital to develop western mineral deposits.

Reading a detailed account of his trip to northwestern Arizona before the National Academy of Sciences in January 1866, Silliman discussed the potential of this region for mineral development. Because these properties held much promise, Silliman examined the practical aspects of mining in northwestern Arizona. Although no trees grew in the immediate area, timber was available within a hundred miles of Fort Mohave and could be brought to the San Francisco District "at a moderate cost" for fuel and other needs. The lack of water might not be a problem, as mining activities in the Comstock Lode of Nevada had produced significant amounts of this resource. Further, wells drilled along dry washes often produced "an abundant supply" of water a few feet from the surface. Agriculture could be developed along the banks of the Colorado River to support a sizable population, although the best lands would have to be obtained from the local Indian tribes through treaty or purchase. Finally, Silliman observed, the climate in the region was generally a healthy one, with "literally no climatic diseases known on that portion of the Colorado." Mining itself would be beneficial in this context as well. Deep mining would offer "refuge" from the high temperatures at the surface and would also provide water of lower temperature and higher quality. Despite various difficulties, the mining districts of northwestern Arizona had great potential.[14]

Silliman's 1864 trip netted him some $40,000 from more than two dozen clients. Later reports and commissions provided him with an additional $30,000 during 1865. This large sum of money (Silliman's Yale salary was less than $3,000 per year) provided a degree of financial security, but failed to balance the harm done to his professional reputation by his association with speculative activity. The revelation that his optimistic report on California oil properties had been based on an analysis of a "salted" petroleum sample troubled him greatly and convinced him to schedule another trip to California in early 1867. He spent the remainder of that year attempting to secure clients and trying to clear up the salted sample question. He was largely unsuccessful in both efforts. His involvement with the earlier California oil speculation limited his marketability and he never disclosed what he had discovered about the questionable sample of 1864. He actively pursued mining interests in the Calaveras area during the spring and summer of 1867, both as a consultant and investor, but this activity led to heavy financial losses for all concerned. When he returned to New Haven in early 1868, Silliman's fortunes were at a low point.[15]

Although he continued his consulting activities, Silliman never again found the financial success he had enjoyed in California in the mid-1860s. He traveled to Utah in the fall of 1871 to examine the Emma Mine south of Salt Lake City. His favorable report led to significant investment in the property through London promoters and encouraged investment in other Utah mines. Silliman returned for a more intensive survey the following February, writing another favorable report that focused on the recent improvements made by the mine owners. Once again, Silliman's reports proved excessively optimistic. By the end of 1872, the main ore body had played out, leading many observers to characterize the Emma Mine

as yet another fraudulent effort to attract investment and raise the price
of stock. Silliman's earlier involvement with the southern California oil
speculation seemed to lend credibility to such charges. Several members
of the National Academy of Sciences, who had been outraged at his earlier
California activity, began a campaign to expel Silliman from the prestigious society in late 1873. Although unsuccessful, this campaign further
damaged Silliman's reputation and became a significant embarrassment.
Rumors continued to circulate about the Emma Mine and Silliman's role in
promoting the venture, ultimately resulting in a congressional investigation. Listening to Silliman's testimony in March of 1876, members of Congress concluded that he had not been an active participant in the Emma
Mine stock swindle. Such a conclusion did little to rehabilitate Silliman's
reputation.[16]

Following the congressional investigation of the Emma Mine, Silliman
attempted to return to his various activities. During the summer of 1877
however, he found himself again involved in the Fort Mohave area, if only
through the print medium. Writing to the *Engineering and Mining Journal*,
published in New York, Silliman announced that he had recently received
a prospectus for the Arizona Chief Gold and Silver Mining Company of
New York. This prospectus included Silliman's description of the Moss
Lode from his 1864 visit, using a fragment of his *American Journal of Science* article to promote the sale of Arizona Chief stock. Silliman, keenly
aware of the potential for fraudulent schemes in western mining, stressed
that there were "important omissions" in the material reprinted in the prospectus. He also pointed out that efforts on the Moss Lode following his
visit had failed to develop a profitable mine. Emphasizing that he knew
nothing about the Arizona Chief except what was included in the prospectus, he vigorously protested "against this unauthorized use of my statements of observations made thirteen years since, and now presumably
cited as a mining report."[17]

Although Silliman remained aware of mining activity in the Southwest,
he did not return to the region until he journeyed to New Mexico in 1880.
The recent arrival of the railroad into the territory had launched a dramatic increase in mining activity and a general economic upturn. Capital
and new technology flowed into New Mexico accompanied by mining engineers and other experts who provided advice and reports. As had been the

case throughout the West, much of the territory's mining had a speculative focus. Consultants such as Silliman continued to play an important role in attracting outside investment.[18]

Silliman arrived in Santa Fe in March 1880, employed by New York mining promoters who had interests in various properties in northern New Mexico. Of initial interest was the Cerrillos region southwest of Santa Fe, which had long been a center of mining activity. Silliman had been hired to investigate the property of the Grand Central Tunnel Mining Company, whose claim had been filed and recorded the previous spring by a group that included territorial Surveyor-General Henry M. Atkinson.[19] Although Silliman's report stressed that only preliminary work had been done on the property, he was cautiously optimistic about future development. Transportation, coal, timber, and other requirements were now available over the Atchison, Topeka, and Santa Fe Railroad, which ran only two and one-half miles from the Grand Central Tunnel claim. A "most equable" climate was an additional advantage, although water was scarce in the immediate vicinity. The geology of the district appeared encouraging and suggested an even greater source of silver veins than the rich strikes in Austin, Nevada.[20]

Silliman's comments were somewhat more cautious in tone than many of his earlier reports on mining properties. He recommended a "careful survey" of the site before expanding operations, but nonetheless advised the claim owners and potential investors that the site had promise. "With proper energy and appliance," he wrote,

> this tunnel ought, after the expenditure of a moderate amount of money, to be made self-supporting, and it is certainly possible, that . . . sufficient ore can be extracted to pay dividends on a fairly large capital, provided always that the enterprise be not starved for the want of money for its proper prosecution. It is not worthwhile to make any estimate of the probable output of this tunnel, as the data for such estimate does not exist. It is sufficient to say, that it is in my opinion a perfectly legitimate and most promising venture, and those who engage in it with a proper sense of its importance, and who are willing to wait its legitimate development, will not be disappointed of their expected reward.

Silliman closed his report by suggesting that the mine might be sufficiently profitable to justify the erection of a smelter to process the ore on site.[21]

The incorporation of the Grand Central Tunnel Mining Company as a New Mexico corporation in July 1881 was at least in part a result of Silliman's generally favorable report. The company, led by Surveyor-General Atkinson as president, issued capital stock of $500,000 in 50,000 shares. The Grand Central company appears to have attracted notable interest, capturing the attention of New Mexico Chief Justice L. Bradford Prince, long interested in land and mining endeavors in the territory. The mine failed to progress beyond the level of speculative activity, however, leaving uncertain the accuracy of Silliman's predictions of future value.[22]

Another property that Silliman had been hired to investigate was the nearby turquoise mine at Mount Chalchihuitl. The site of previous mining activity, this mine had recently been reopened by D. C. Hyde, who had founded the Grand Reserve Consolidated Gold and Silver Mining Company in January 1880. Silliman published four articles on his investigations, including an extensive discussion in *American Journal of Science* and shorter accounts in *Science*, the *Engineering and Mining Journal*, and the *Proceedings* of the American Association for the Advancement of Science. These papers stressed the likelihood of earlier Spanish mining activity and the supposed connection between gold and turquoise deposits. Hyde soon changed the name of his mining company to the Turquoise Gold and Silver Mining Company and was moderately successful in attracting investors. He was completely unsuccessful in his efforts to find gold, however, and no record of his company exists after 1882.[23]

Silliman's somewhat cautious perspective on these New Mexico properties did not extend to his third assignment in the territory. Accompanying one of his clients (New York Stock Exchange member J. M. Seymour) to the Taos region in April, Silliman examined potential gold deposits in the Rio Grande Valley. His report, published as a thirty-four-page pamphlet at the end of the month, enthusiastically recommended the use of hydraulic mining techniques to recover the valuable mineral. In the introductory material to his report, Silliman congratulated the investors who controlled these properties (including Surveyor-General Atkinson) for their "sagacity in amassing this gravel area" and described the gold-bearing region as "on a scale of magnitude surpassing anything since the discovery of California."

Silliman provided a detailed description of the geology of the region, explaining the necessity of hydraulic mining to recover the gold in the gravel deposits. Such a technique would require significant capital, but the potential for success was particularly great. These gravel beds were much thicker than those that had been successfully worked in California, leading Silliman to predict that the yield per cubic yard would probably be greater in New Mexico than in California. Once improvements to the New Mexico properties had been made, the engineering techniques that had been so successful in California could be applied to the Rio Grande gravels with equal promise of success.[24]

Silliman's enthusiasm for this project knew few bounds. He told his readers, "Nothing, I am persuaded, since the discovery of California and Australia, is comparable for its measurable reserves of gold, available by the hydraulic process, in these deep placers of the Rio Grande." At the end of his report, having described the techniques required and the potential for success, Silliman offered a final word of encouragement: "With proper attention to these details, the use of the best experience, and the economical employment of ample funds, the owners of this magnificent property will reap a rich and long continued reward from their New Eldorado." Silliman's recommendations were never implemented, however, and the Rio Grande gold gravels remained undeveloped except as a speculative venture.[25]

Following his visit to northern New Mexico, Silliman returned to New Haven and other consulting activities in the East. During the fall, following an extensive excursion in the mountains of western Pennsylvania, Silliman began to display symptoms of heart disease that left him prostrated for several weeks. His recovery progressed sufficiently, however, that by the late summer of the following year he was preparing for another trip to New Mexico, this time focusing on the southern portion of the territory.[26]

Silliman left New Haven in late September, following publication of a notice in the *Engineering and Mining Journal* that he would be traveling in the Southwest "and will attend to any commissions on his line of travel." The mining business in New Mexico continued to attract the interest of eastern capitalists and mining experts. Observers reported the great potential of the territory's mineral deposits, noting that several companies were working mines in Socorro, Grant, and Dona Ana counties in the southwestern portion of New Mexico. The Shakespeare District in Grant County, for example, had attracted companies from Indiana and Missouri, and had produced sufficient gold ore to convince investors to finance the construction of a smelter.[27]

Traveling south from Santa Fe on the recently completed rail line, Silliman began his examination of southern New Mexico in Socorro, the center of the region's mining interest. In a later published discussion of his survey, he noted the area's immediate past. "It is only very recently," he wrote, "that large portions of this region have been accessible. It has been full of danger from hostile bands of wandering Apaches, whose murderous raids have slain many adventurous miners and seriously prevented developments. But this evil has now been largely abated, and in many of the mining camps of this region there is no longer any danger from this cause." Silliman was thus able to begin his investigations in relative safety, visiting several mines west of Socorro during his October sojourn. The mines in this area, extending into the Magdalena Mountains some thirty miles west of the city, were promising silver strikes, showing sufficient ore to justify the erection of stamp mills and other improvements.[28]

Of much greater interest to Silliman were the numerous mining endeavors in the Black Range of Socorro and Grant counties, especially those in Lake Valley (after 1884 in the newly created Sierra County). A significant body of silver ore had been discovered in the Lake Valley area in the late 1870s, followed by the creation of several companies to control the best claims. When Silliman visited these properties in mid-October 1881 he did so with the encouragement of well-known mining promoters George D. Roberts and J. Whitaker Wright, both of whom were involved in speculative activity in the area.[29]

Silliman's examination of Lake Valley revealed much of interest. The deposits controlled by the leading mining companies yielded silver ore "in greater or less abundance" and could be readily developed because of recent transportation improvements and a congenial climate. Assays done on site by Silliman and by the New York Metallurgical Works justified continued interest and investment. In fact, Silliman concluded, the Lake Valley deposits appeared very similar to those of Leadville, Colorado, and

Eureka, Nevada, and were likely to be as valuable and permanent as these famous silver strikes. Having completed his survey of the region's mineral prospects, Silliman returned to New Haven in early November.[30]

Over the next few months, activity in the Black Range increased dramatically, especially after the discovery of the rich silver deposit known as the "Bridal Chamber" shortly after Silliman's departure. A geologist who wrote about the region for the scientific journal *American Naturalist* observed, "large numbers of prospectors, miners, mechanics and a fair percentage of professional scoundrels have flocked into the country."[31] In order to gather additional information concerning the region, Silliman visited New Mexico again during June of 1882, attracting much interest from the New York *Tribune* and the *Engineering and Mining Journal*. Silliman's party, which included several Eastern investors and members of the Philadelphia syndicate that controlled the major claims in Lake Valley, were accompanied by a military escort to protect them from both Indians and outlaws, a precaution that proved unnecessary. Silliman noted the continued promising production of ore and emphasized the improvements since his last visit. At Lake Valley, a recently completed twenty-stamp mill was processing as much as 100 tons per day, attracting the attention of Silliman and his party as well as that of the *Tribune* reporter supplying special dispatches from New Mexico.[32]

Based on his two visits, Silliman's report on the Lake Valley mines was a positive one, reinforced by shipments of bullion from the area to Philadelphia during the summer of 1882. Yet the speculative nature of these properties and the call for investment based on Silliman's report raised questions at the *Engineering and Mining Journal*. Editors Richard P. Rothwell and Rossiter W. Raymond, both of whom were experienced mining engineers, identified the leading figure in Lake Valley as George D. Roberts, "whom the investing public should thoroughly appreciate as a manipulator of mines." Equally troublesome was Silliman's involvement in the current promotion of the mines as investment properties. Acknowledging Silliman's significant standing as a scientist, the *Journal* pointed out that his record as an expert in the examination of mines "is such as to make it necessary to receive his estimates with extreme caution." In the past, the periodical observed, he "has displayed a lack of judgment which should suffice to render impossible any attempt to float mining property on his unsupported testimony." The combination of Silliman's past record and the fact that Roberts was clearly in charge of the Lake Valley scheme led the *Journal* to conclude, "we must warn investors not to touch the stock."[33]

The speculative nature of Lake Valley developments soon moderated. Silliman presented a paper on his journeys to New Mexico at the October meeting of the American Institute of Mining Engineers in Washington D.C. Perhaps responding to the criticism of Raymond, Rothwell, and others, Silliman offered a cautionary note in the closing sections of his address. He told his audience that many additional positive comments could be made concerning various mines in the Black Range area, "but it is enough to add . . . that systematic exploration, in depth, alone will develop its hidden treasures, and not the manipulation of its shares in Eastern cities." A month later, the *Engineering and Mining Journal* reported that Roberts had passed control of the Lake Valley mines to "the Philadelphia party" who had initiated a new spirit into the enterprise. A new smelter and the expansion of production were expected soon. Mining activity in the region remained notable throughout the 1880s (Lake Valley mines produced an average of $100,000 in bullion per month during 1883) but declined rapidly thereafter.[34]

Silliman's return to New Haven during the summer of 1882 marked the end of his western adventures. He never again traveled into the Southwest, but he remained interested in the region. He continued to learn of developments in Arizona, often through his friend George A. Treadwell, a mining promoter whom he had met during the Emma Mine controversy. Beginning in the mid-1870s, Treadwell kept Silliman informed of western mining developments and often sent the Yale chemist samples from mines of interest. He also suggested various properties to Silliman as investments, including mines on the Comstock Lode and in Arizona.[35]

At about the time Silliman began his investigations of New Mexico properties in 1880, Arizona entered its own dramatic mining boom, following discoveries in Tombstone and elsewhere. Treadwell alerted Silliman to these developments, but also continued his efforts to secure good mining investments for his Yale friend. Potential copper mines in Arizona and Sonora were the topics of letters sent to Silliman, while Treadwell's brother, John, sought to involve him in a possible gold property near Fort Tejon in California. John asked Silliman to advise one of the principals in the Snowy Mining District holdings who hoped to sell the property. If the sale went through, Treadwell wrote, he would "see that you make from 2500$ to five thousand dollars coin. This is only between you and me. All moneys will pass through my hands if a sale is made. This is not to be known." Although nothing came of any of these ventures, George Treadwell continued to encourage Silliman to remain active in supplying his expertise as a mining consultant. Such work was valuable and worth generous fees. Even more important, Treadwell reassured him, "I will find a mine yet that will pay us well."[36]

Among the areas that came to Silliman's attention were the silver deposits in Yuma County. Claims had been filed in the region as early as the 1860s, but it was not until the late 1870s that noteworthy production began in the Silver District and Castle Dome District. These deposits attracted the attention of recently appointed territorial governor John C. Frémont, who devoted much of his term in office to speculative activities. During several trips to New York during 1879 and 1880, Frémont remained very active in the promotion and sale of Arizona mining properties, including those in Yuma County. His most profitable activity came in May 1880, when he sold the Red Cloud Mine in the Silver District (in which he held a substantial interest) to a New York syndicate. During this eastern trip, Frémont and Silliman became acquainted. Although the two did not collaborate in any of the governor's promotional schemes, Frémont attempted to keep informed of Silliman's activities through their mutual friend George Treadwell.[37]

The change in Red Cloud ownership led to increased activity in the Silver District by early 1881. Silliman's reputation as a mining consultant led several miners to send him samples and information concerning their properties. Although the chemist undoubtedly provided these miners with advice, he also published his findings in the scientific press. His analysis of samples from the Silver District, for example, appeared as an article in the September 1881 issue of *American Journal of Science*. Discussing the Red Cloud and other mines, Silliman stressed that the rock of the district was very similar to deposits in Nevada and represented "the usual associates of silver ores the world over." Famous mining engineer William P. Blake had examined the Silver District a few months earlier and echoed his colleague's favorable opinion of the region. Despite these evaluations, the Red Cloud Mine failed to live up to its promise and was sold at a sheriff's auction in October 1882. Other mines in the area were no more successful, with the low grade ore insufficient to justify the expense of development.[38]

Although Treadwell was interested in mining activity throughout Arizona Territory, his focus remained the Wickenburg and Prescott areas, long noted as rich mining regions. During Silliman's 1880 trip to New Mexico, Treadwell sent him a package of samples from the Phoenix Mine near Cave Creek. Treadwell soon purchased the mine for J. M. Seymour, who had accompanied Silliman on his northern New Mexico trip. Although Silliman's assay yielded indifferent results, the mine nonetheless proved profitable and provided shareholders with a reasonable return on their investment.[39]

Silliman's September 1881 *American Journal of Science* article included numerous other comments concerning Arizona mining. Writing about the famous Vulture District near Wickenburg, he described the mines in the area and their mineralogical characteristics. This article attracted much attention. Treadwell's friend, Thomas F. Hopkins, editor of the *Phoenix Herald*, planned to reprint the article in his newspaper. From San Francisco, the *Mining and Scientific Press* discussed Silliman's work in late October. The Phoenix Mine continued to attract interest as well. By the end of the year, the *Engineering and Mining Journal* told readers of the recent purchase of the property by eastern capitalists and the expansion of the mine to include a 100-stamp mill.[40]

Encouraged by such developments, Treadwell continued to send Silliman samples from mines in the region and assured his friend that he was still looking for investment and other financial opportunities. In the fall of 1881 he wrote that he would "keep a good eye out for you if I see any chance," followed in March 1882 by a note that he was attempting to arrange for Silliman to conduct some mineral analyses for a generous fee. The Red Rover Mine, some sixty miles north of Phoenix, attracted Treadwell's interest in late 1882. Silliman analyzed samples sent by Treadwell and apparently received stock in the company early in 1883. Although this mine displayed rich silver deposits near the surface it proved of greater value for its copper deposits which were developed much later.[41]

Because Treadwell was heavily involved in the developments that led to the emergence of copper as the centerpiece of Arizona mining, Silliman's involvement in the copper industry is hardly surprising. Although copper claims had been filed in what would later become the Jerome area as early as the mid-1870s, significant attention in the area did not arise until 1880. Silliman's involvement began in November of that year, when he received a long letter from C. P. Head and Hugo Richards concerning the Eureka Mine. Treadwell had suggested that these two promoters write to Silliman as part of their (and Treadwell's) efforts to attract eastern investors. During the next month, Silliman used his contacts in the East in an effort to market the Eureka Mine and played an important role in arranging for the young James Douglas, later the key figure in the emergence of Phelps Dodge, to visit the area in December. Douglas visited the Eureka Mine and the Wade Hampton claim on behalf of potential Philadelphia investors and, despite the long distance to a rail line, urged that they exercise their option on these claims. By February of the following year, the Eureka situation looked quite different. According to Treadwell, Richards had been less than forthright in some of the information he had supplied, leading Douglas to file a more optimistic report than was perhaps justified. Treadwell was especially upset with Richards's dishonesty, as Silliman had been caught in the middle and had nothing to show for his effort.[42]

By the spring of 1883, however, further exploration of the claims revealed that the copper deposits in the area were in fact very rich. Treadwell had taken a leading role in promoting these properties, convincing recently appointed territorial governor Frederick A. Tritle and New York investor Eugene Jerome to become actively involved. Treadwell, Tritle, Jerome and other investors soon formed the United Verde Copper Company to develop the property. Although marginally profitable operation characterized the company's first few years, transportation difficulties retarded the development of these claims and eventually led to the purchase of the property by copper baron William A. Clark in 1889. Despite his relationship with Treadwell, Silliman's inability to invest in the Eureka Mine during 1880 and 1881 left him no connection to the growing copper boom in Arizona.[43] Later involvement with another Treadwell effort in Copper Basin southwest of Prescott proved no more profitable, although this time Silliman apparently received a large amount of stock. Unfortunately, the mine proved less than marketable and by early 1884 the principals had all but given up selling the property.[44]

After the collapse of his copper mining investments in early 1884, Silliman's activities focused on scientific pursuits in New Haven. In September, for example, he presented a paper at the Philadelphia meeting of the American Institute of Mining Engineers, offering a technical discussion of the use of natural gas in recently improved glass furnaces. This would be his last scientific work. The following month, the heart condition that had troubled him earlier in the decade became especially pronounced and was soon complicated by pneumonia. The combination precipitated a slow decline in his health until his death on 14 January 1885.[45]

His role as a scientific consultant, especially during the post–Civil War period, in many ways defined Silliman's status in the American scientific establishment. The dramatic increase in speculative activity during America's Gilded Age offered many opportunities for scientific consultants such as Silliman, especially in the mineral industries. The commitment to scientific disinterestedness, which underlay the concept of scientific expertise, became increasingly relaxed during the economic expansion of the late nineteenth century. Because of his personality and outlook, Silliman appears to have been an excellent example of the new consulting attitude. As the leading historian of scientific consulting wrote, "Silliman was optimistic, outgoing, and trusted others (scientists and businessmen) almost to a fault; as a result, he was perhaps the most relaxed of these consultants with regard to ethical standards."[46] The Yale chemist thus found himself criticized by his colleagues, especially when mining properties he had praised highly proved less than successful. Although his role in the California oil fiasco and his association with the Emma Mine scandal were the most dramatic (and damaging) examples of his consulting activity, his involvement with numerous developments in Arizona and New Mexico often displayed similar characteristics.

Nonetheless, the contributions of consultants to the growth of the Southwest's mineral wealth were significant. Without such expertise, investors would have had little reason to commit the large sums of money necessary to develop successful mineral deposits. Although many of these ventures failed, the region's economic growth in the late nineteenth century was in large part a result of those mining ventures that proved successful. Silliman's role in the growth of the Southwest's mineral economy, although focused on speculation rather than production, provides valuable insight concerning the region's development in the late nineteenth century. His role also suggests, however, that scientists involved in this development often found themselves navigating through an ambiguous terrain.

Chapter 3 Tree Rings and Climatic Cycles: A. E. Douglass

The role of Andrew Ellicott Douglass (1867–1962) in the development of science is well known. His service with the Lowell Observatory helped to establish that facility and alerted astronomers to the superiority of southwestern sites for telescopes. As director of the Steward Observatory at the University of Arizona, Douglass guided another important institution and laid the foundation for the emergence of southern Arizona as a center for visual astronomy. He is perhaps most famous for the development of dendrochronology and its dramatic application to archaeology. Through the analysis of the annual growth rings of trees, Douglass provided archaeologists with a precise chronology for the ancient ruins of the Southwest. Yet Douglass's tree ring studies were, in fact, an outgrowth of his research in astronomy. Further, he soon recognized the potential value of this work for climatic studies. His efforts to employ tree ring analysis as a method for long-range weather forecasting increasingly defined his activities and remained the aspect of his endeavors that Douglass believed would be most important.[1]

Born in Windsor, Vermont, Douglass graduated from Trinity College in Hartford, Connecticut, in 1889. He served on the Harvard College Observatory staff until 1894, when he joined the newly established Lowell

Observatory. He traveled to Arizona Territory during the spring, located the site for the facility on a mesa west of Flagstaff and coordinated the construction of the observatory. He played an important role in the study of Mars that defined Percival Lowell's career and served as the facility's principal assistant for the next seven years. Douglass's criticisms of Lowell's theory of intelligent Martian life led to his dismissal in 1901. The young astronomer remained in Flagstaff where he had established a wide circle of friends, serving variously as an assayer, teacher in the territorial normal school, and probate judge. In 1906, however, he secured an appointment to the faculty of the University of Arizona in Tucson. Although a small school with minimal resources, the university offered Douglass an opportunity to return to science.[2]

By the time of his relocation to Tucson, Douglass had already taken initial steps toward a new scientific career. Following his dismissal from the Lowell Observatory, Douglass journeyed throughout the region and noticed the effects of a particularly dry summer on the pine forests of the region. He also observed that rainfall in the region varied directly with altitude and assumed that rainfall controlled tree growth. The next step in Douglass's chain of reasoning proved crucial. As he recalled nearly forty years later, his reflections on the forest growth of Arizona led back to astronomy. "If the altitude controls the rain and so the trees," he recorded in a manuscript from 1939, "why shouldn't the sun in its changes from year to year control the rain and so control the trees; and the annual rings of the trees should show it. Hence why not measure the successive rings of trees to see if their growth from year to year would show a sunspot cycle."[3]

In his initial investigations, Douglass benefited from significant research on solar phenomena. The eleven-year sunspot cycle had been studied and analyzed for more than a half century and had been correlated with similar cycles in terrestrial magnetism and auroral displays. Possible connections between solar and terrestrial phenomena, especially the link between sunspots and Earth's climate, also attracted interest. Douglass perused studies of tree growth, weather cycles, and solar variation in the *Astrophysical Journal* and the *Monthly Weather Review*, gaining valuable knowledge of these topics.[4] Recognizing that research concerning the solar-terrestrial link remained at a preliminary level, he had few qualms about preparing a discussion of his own efforts for publication.

Completed during the fall of 1908, Douglass's contribution appeared as "Weather Cycles in the Growth of Big Trees" in *Monthly Weather Review* for June 1909. Following a description of Arizona's climate and a discussion of the pine forests near Flagstaff that provided his tree ring record, Douglass discussed the potential value of such trees as record-keeping devices for climatic changes. The arid climate of Arizona made the region's trees especially sensitive to rainfall variations, thus providing Douglass with an accurate precipitation record. The twenty-five specimens of yellow pine he had collected, with their average age of 324 years, yielded nearly 10,000 ring measurements. This record allowed Douglass to trace climatic fluctuations over time and to employ tree ring records as valuable instruments for both climatic and astronomical research.

Of far greater interest to Douglass, however, were possible cyclical variations in the ring patterns of his Flagstaff trees. Four such cycles appeared particularly intriguing. A short period of nearly six years seemed to emerge in pairs, giving a full period of between eleven and twelve years. A period of slightly more than eleven years appeared separately, and struck Douglass as indicative of a sunspot cycle parallel. Longer cycles of nearly twenty-one and thirty-three years were also visible in the ring record. Douglass found similar cyclical variations in temperature and precipitation records from San Francisco, San Diego, and Sacramento. Carefully studying the curves of temperature, sunspot activity, and tree growth, Douglass found "unmistakable" evidence of the eleven-year period associated with the sunspot cycle. That the factors involved in sunspot and other solar activity likely had a "profound influence" on terrestrial conditions made the tree ring record of immense potential value for a better understanding of terrestrial climate.[5]

As his scarce time permitted over the next few years, Douglass continued to expand his tree ring collection and to compare the ring patterns with weather records. By 1912, he had determined that characteristic patterns of wide and narrow rings appeared in numerous specimens gathered in both the Prescott and Flagstaff areas. As he developed the technique of cross-identification (or cross-dating, as it later became known) Douglass was able to extend his tree ring chronology more accurately. Specific patterns would provide a date for the tree specimen even if a cutting date was unavailable. Further, the assignment of precise dates to his tree ring collection allowed him to refine certain ideas advanced in his *Monthly*

Weather Review article. The tree curve presented in this publication, Douglass later remarked, had not shown the eleven-year period as closely as he had wished. This situation had at least in part been a result of the lack of certainty regarding precise dates of characteristic rings. Once cross-identification had been established, however, Douglass reworked these data to test their association with the sunspot period "and far better results were obtained."[6]

Douglass was well aware that his data from sunspot and climate records, as well as his tree ring specimens, failed to show precise mathematical fluctuations. He never expected such a result, as even the sunspot record showed variations in the value of the cycle over time. Douglass was, in short, focusing on more general patterns of variation. To pursue the analysis of such trends Douglass developed a mechanical device to identify and measure periodic fluctuations. First constructed in the spring of 1913 at Harvard College Observatory, Douglass's "periodograph" (later called a "cyclograph" or "cycloscope") produced a photographic image that revealed patterns in the data analyzed. Douglass constructed a curve of the data (initially the sunspot record) on white paper cut out and placed against a black background. He created several copies of this image, mounting each additional curve ten years to the left and below the previous one. By rotating this multiple plot and exposing it photographically through an appropriate lens, Douglass produced an image in which specific periods (if they existed) were displayed as recognizable patterns. This technique provided a rapid method to analyze periods of various values as the resulting "periodogram" eliminated the "noise" in the data.

Douglass described his new device and its applications in two *Astrophysical Journal* articles. Analyzing the sunspot numbers since 1750 and the rainfall records from Windsor, Vermont, between 1835 and 1912, Douglass noted both the cyclical nature of the data and the discontinuities that other observers had noted. Douglass's photographic images showed a 9.3-year sunspot cycle in the late 1700s, followed by a thirteen-year cycle between 1800 and 1830. After this latter "interval of readjustment," the 11.1-year sunspot cycle had remained in evidence since 1830. Expectedly less regular than the sunspot cycle, the Windsor rainfall record nonetheless revealed a mean period of some twenty-eight months, despite notable inconstancy. This analysis paralleled other climatic studies of the late nineteenth century and convinced Douglass that his equipment was reliable. More importantly, however, the analysis of the Vermont rainfall record demonstrated the value of taking a broader look at periodic fluctuations in data. As he wrote in 1915, "Hence we find some cause for believing in a period which is not constant. In applying rigid periods this is likely to be overlooked. The possibility inferred from these photographic results then is that the study of slightly variable periods might open up added fields of knowledge of meteorological phenomena. To the investigation of such periods, the instrument here described is especially adapted." Douglass continued to employ modified versions of this equipment throughout his long career.[7]

Douglass's first extensive published discussion of the new science of dendrochronology appeared in late 1919 through the auspices of the Carnegie Institution of Washington. *Climatic Cycles and Tree-Growth: A Study of the Annual Rings of Trees in Relation to Climate and Solar Activity* provided a succinct summary of the ideas and techniques behind Douglass's research and described the data obtained from the more than 75,000 individual rings measured. As indicated by the title, however, Douglass's primary goal was to present evidence for cyclical climatic variations and their relationship to solar activity. Throughout *Climatic Cycles* he emphasized the connection between tree growth curves and the cyclical variation of sunspot activity, observing that many specimens in his collection followed the sunspot curve "almost perfectly." Douglass also noted that solar physicists had provided evidence that the sunspot cycle appeared to vary over time, attempting to make clear the distinction between precise mathematical periods of variation and "cyclic" patterns such as those he studied.[8]

Although temperature, precipitation, and tree growth curves showed a "marked relationship" with the sunspot curve, Douglass's data actually revealed a more complex situation. A number of cycles appeared in the tree ring record, but the "more conspicuous and general cycles at once apparent" were all "directly related to the solar period." Douglass had isolated a half sunspot period of five or six years, the sunspot period which varied between ten and thirteen years, and cycles which indicated multiples of this latter value. He described these as a "double" sunspot period (21 to 24 years), a "triple" period (32 to 35 years), and a "triple-triple" period of 100 to 105 years. Douglass acknowledged that the analysis of

A. E. Douglass as he appeared in the early 1920s. (Courtesy of University of Arizona Library Special Collections; A. E. Douglass Papers, AZ 72, box 156, folder 2A)

such cyclical variation was in its initial phase, but the potential for long-range prediction of climate based on such variation, as well as the potential for the study of solar activity in the past through the analysis of tree ring records, clearly justified the expansion of dendrochronological research.[9]

As evident from *Climatic Cycles and Tree-Growth*, the concept of "cycles" was increasingly guiding Douglass's research and remained a key interest of the Carnegie Institution of Washington. Organizing the Conference on the Origin and Nature of Cycles in Washington on 8–9 December 1922, the institution invited several investigators to examine such topics as solar variation, economic and health cycles, and the analysis of cyclical phenomena. Douglass's presence at this conference was deemed "indispensable" by Carnegie official F. E. Clements, who telegraphed the Arizona astronomer that the institution would pay his travel expenses between Tucson and Washington. Douglass would join such figures as Ellsworth Huntington of Yale, Charles Greeley Abbot of the Smithsonian Institution, and Walter S. Adams of the Mount Wilson Solar Observatory, among others.

Clements opened the conference with a discussion of the "Nature of the Problem of the Cycle," emphasizing the difference between a cycle of recurrence which could be measured even if its actual value varied, and a specific periodicity of an invariable quantity. The sunspot cycle was an excellent example of the former, as were long-term climatic variations. Because Douglass's work clearly focused on such cycles he gave the opening paper for the conference, summarizing his analytical methods and the conclusions he reached concerning cyclical variation. His discussion of cycles again stressed the eleven-year sunspot cycle and its multiples, although clearly within the context of cyclical variations rather than precise periodicities. During the discussion that followed Douglass's and other papers, participants generally agreed that the cycle concept did not necessarily mean a precise, mathematical relationship and should not be expected to lead to equations relating various phenomena to each other.[10]

In retrospect, Douglass's participation at the Carnegie cycle conference may be viewed as a watershed in the history of dendrochronology. Although he never abandoned the sunspot cycle as an important theme in his work, Douglass increasingly followed two other research paths. The best known aspect of Douglass's later research and the activity that dramatized the value of his new science was the application of tree ring research to archaeology.[11] Despite his archaeological contributions he continued to believe that dendrochronology had similar potential for long-range weather forecasting. To alert his scientific colleagues to this potential, Douglass began work on a second volume of *Climatic Cycles and Tree-Growth* focusing more specifically on cyclical concepts. He completed an outline of the proposed work in the fall of 1923. This second book would provide more studies of cycles, using material from the Southwest and New England, and would examine such topics as business cycles, geographic effects on cycles, clay layer analysis, and general cycle theory. The Carnegie Institution proved receptive to Douglass's plan and provided financial support, but preparation of the manuscript proved much more involved than Douglass or his Carnegie colleagues anticipated. Completion of the volume required three years before its publication in mid-February 1928.[12]

The second volume of Douglass's study clearly showed a new focus. Although he included various discussions about his tree ring studies and the analysis of data through his cyclograph methods, Douglass emphasized cyclical variations in the tree ring record. Sequoias, for example, showed an eleven-year cycle on many occasions since 1300 B.C. Even more interesting, the large collection of sequoia records from the past five hundred years showed several additional cycles of ten, fourteen, and twenty years, the latter of which was particularly prominent. The long Flagstaff record from 1300 to 1925, however, provided Douglass with the most valuable evidence of cyclical variations. The pines consistently displayed evidence of the sunspot cycle, but also indicated cycles of seven, fourteen and twenty-one years beginning approximately 1660. Climatic cycles were similarly evident. Dry years as shown in the Flagstaff pines suggested cycles of fourteen and twenty-one years, with major droughts at intervals of approximately 150 years. Minor droughts appeared to occur at forty- to fifty-year intervals. Similar cycles appeared in pines gathered elsewhere in the West, leading Douglass to conclude that the cycles displayed in these records appeared to be simple fractions of a thirty-four to thirty-five-year cycle. The eleven-year cycle remained the most evident and intriguing for Douglass, as it represented a one-third value of the proposed longer cycle.[13]

Despite the suggestive and intriguing material provided by his tree ring records, Douglass acknowledged that his work remained in its early stages. The concept of cycles of sufficient reliability to lead to prediction required much further thought. "Until we know the physical cause of cycles," Douglass wrote, "we can not say how long a mechanical repetition will last, for it may break down at any time." The Flagstaff record, for example, showed the eleven-year sunspot cycle for centuries, but it faded in the mid-seventeenth century and was replaced by cycles of other values. Nonetheless, Douglass remained convinced that the continued analysis of cyclical changes in the tree ring record would lead to a better understanding of climatic variations.[14]

The Carnegie Institution remained committed to the study of climatic cycles and organized a second conference for mid-December 1928. Chaired by noted botanist Daniel T. MacDougal, the conference focused more clearly on climatic concerns and emphasized the correlation of MacDougal's work on the physiology of tree growth with Douglass's research on tree ring phenomena. In his introductory remarks, Carnegie Institution President John C. Merriam emphasized the value of cycle studies to the understanding of climatic variation and the importance of tree growth studies to this understanding. Following MacDougal's discussion of the physiology of tree growth, Douglass presented his paper, "Cycles in Tree Growth: The Cyclograph," in which he described the instrument that remained the center of his analytical endeavors. He also summarized the various cycles he had discovered in his western tree specimens, showing the influence of rainfall on tree growth and the link between these cycles and solar variation. The thirty-five scientists who attended the conference discussed various other cyclical phenomena related to climate that suggested further research. All agreed, however, that the pursuit of cycle studies offered much of value.[15]

As Douglass redirected his efforts toward the study of climatic cycles as revealed in the tree ring record, his research attracted the attention of other scientists interested in climatic variation. In early 1932, he received an invitation to present a paper at the April meeting of the National Academy of Sciences in Washington, D.C. The academy's Committee on Long-Range Weather Forecasting was organizing a symposium on climatic cycles and was eager for Douglass to participate. The symposium would include discussions of other cyclical variations related to climate, such as solar radiation and terrestrial temperature, sunspot cycles, and possible cyclical variation in such phenomena as glaciation and streambed deposits. Douglass's paper, "Evidence of Cycles in Tree-ring Records," would open the program.[16]

In his presentation to the academy, Douglass attempted both to describe his methodology in a convincing manner and to discuss the preliminary results of his research. The long tree ring record of some three thousand years not only provided an accurately dated chronological record but also supplied a climatic record based on ecological considerations. Equally important, Douglass's cyclogram method of analyzing plotted curves supplied a method uniquely suited to the "cyclic" data from tree rings and weather records. Realizing the importance of establishing his cycle concept as a legitimate one, Douglass discussed this aspect of his research at some length. The sunspot curve had been an important part of Douglass's tree ring research for more than three decades and had been a research topic for various astronomers for nearly a century. The sunspot record, however, failed to present a mathematically precise and permanent curve. "It is evident that the solar cycle is not a precise and invariable period," Douglass announced, "yet we call it a cycle. Therefore a cycle may be defined as the recurrence of similar conditions at similar intervals." Such cycles were those found in climatic records, including tree ring patterns, and could best be analyzed through techniques that did not rely on mathematical equations. Douglass's cyclogram method was one such technique.

Douglass devoted the remainder of his address to a description of his results. Cyclogram analysis revealed evident cyclical patterns, most of which were "very nearly simple fractions of small multiples of the sunspot cycle." The sequoia record, in contrast, revealed a double sunspot cycle of between twenty-two and twenty-three years, although these ancient trees also displayed longer cycles of note. Douglass mentioned a sequoia "recurrence cycle" of approximately 275 years and two other cycles near 100 and 220 years in length. The Arizona astronomer acknowledged the preliminary and complex nature of his cycle analysis, but he remained convinced of the potential value of these studies. Especially from the long sequoia record, Douglass concluded, students of climatic phenomena might well "obtain in some form of cycle succession a real step forward toward a theory of climatic change and long range prediction."[17]

Following the other papers, four academy members presented formal comments. Each found the various cycles discussed in the papers intriguing, but the concept of "cycle" itself proved the topic of greatest interest. Noted electrical engineer Arthur E. Kennelly, who had been an assistant to Thomas Edison and who had recently retired from the Harvard faculty, found much of value in the papers. He admitted that definite conclusions concerning the relation between solar and terrestrial phenomena remained premature, but the correlations discussed at the academy symposium suggested that more extended studies would be worthwhile. Statistician Edwin B. Wilson of the Harvard School of Public Health also encouraged further investigation of cycles and urged the examination of additional variables in such studies.[18]

Yale mathematician Ernest W. Brown, best known for his work in celestial mechanics, was less impressed. His mathematical perspective convinced him that the "cyclic" relationships described by the speakers were questionable. Mathematically described cycles were the only ones of value and the only ones that might lead to the discovery of the underlying causes of the variation. Even more suspect was the concept that the phenomena under study could lead to the discovery of several cycles. Clearly challenging Douglass's work, Brown argued that, "Where several cycles are found, the laws of chance alone may give us one in one set which nearly corresponds with one in the other set. This is particularly the case with the sub-harmonics. If we find a cycle for example, of twelve years, we are almost sure to find one or more of six, four, three, etc., years." The only true climatic cycles, Brown concluded, were the day and the year, both of which represented geometrical changes in Earth's position with regard to the Sun.[19]

Similarly critical of the cyclic concept was the eminent Princeton astronomer Henry Norris Russell. He opened his comments by pointing out that solar and especially weather periodicities were quite different from those familiar to astronomers. The sunspot cycle, he argued, was the result of a release of some sort of internal strain. Therefore, each sunspot cycle was a unit. "One would not expect such cycles to be more than very roughly periodic," Russell observed, "like the eruptions of one of the more regular geysers." Russell remained skeptical that weather prediction based on solar and tree ring studies would lead to the desired goal, as he stated near the end of his comments. "To disentangle the various cycles from one another and from the other fluctuations," he concluded, "is a task of excessive difficulty, and I fear that even generalized prediction of weather conditions by the extrapolation of empirical cycles derived from observation, if attainable, is far in the future." Russell nonetheless encouraged the continuation of these studies and emphasized that the research had led to valuable knowledge. Students of solar radiation had shown the fluctuation of this phenomenon, with potential value to astrophysicists. Even Douglass's work, which Russell found of less value, "has provided a precise chronology of fundamental value to the western archaeologist."[20]

The critical comments from Brown and Russell discouraged Douglass. Having designed his presentation to show that the data under analysis clearly did *not* display the characteristics of regularly variable values such as a sine curve, he found his research challenged for that very reason. During the summer and fall Douglass sent letters to several scientists in an effort to explain his concept of cycles and their analysis. In early November he drafted an analysis of his difficulties to his friend and Carnegie colleague Herman A. Spoehr. Although never mailed, the letter provides insight concerning Douglass's outlook. "I don't understand the opposition of such men as Russell and Brown," Douglass wrote. "It can be due to nothing else than that they do not know what I have actually done. I don't know how to get it over to them." It was clear at the April symposium, he told his friend, "that they did not have a ghost of an idea of what I was doing or what my method was."[21]

Douglass would have another opportunity to present his ideas at a Carnegie Institution exhibit and lecture in early December. Throughout his preparation, which consumed most of his time during the second half of 1932, Douglass followed the suggestions of Carnegie officials Merriam and Spoehr, who stated that Douglass's work would clearly be on trial in December. The exhibit of tree ring specimens and his lecture were equally important and both needed to be crafted carefully. Douglass focused on organizing his material to show the correlation between solar activity and tree growth, but also planned to stress the distinction between the mathematical cycles studied by astronomers and the less regular but nonetheless cyclical variations displayed in solar and climatic records.[22]

During his lecture, Douglass emphasized the impressive amount of data that underlay his work. Not only had he analyzed more than 250,000 tree rings, but he had also examined other climatic records such as clay varves and mineral deposits which showed evidence of the sunspot cycle. Douglass was quick to point out, however, that the sunspot cycle did not always reach its maximum value at precise eleven-year intervals. One of the most dramatic departures from the standard sunspot frequency occurred during a "dearth" period of sunspot activity in the late seventeenth and early eighteenth century, when the cycle actually displayed a value closer to ten years. Minor cycles displaying other frequencies also appeared in the tree ring record, providing Douglass with an opportunity to describe his cyclogram method of analysis. This technique allowed the researcher to eliminate the minor cycles so that the underlying main cycles became clearly distinguishable. By carefully describing both his data and his method of analysis, which was uniquely suited to the tree ring and sunspot records, Douglass hoped to convince his scientific colleagues of the merit of his research program.[23]

Despite Douglass's careful preparation and the support of Carnegie officials, the December conference was no more successful than the one in April. Critics continued to challenge the concept of cycles that could not be described by equations and remained skeptical of the existence of multiple cycle values. In addition, Douglass's tree ring methods themselves came under criticism. The technique of cross-dating, which allowed Douglass to build long chronologies based on characteristic patterns of wide and narrow rings, struck some eastern dendrologists as questionable. Unfamiliar with the sensitivity shown by trees in arid climates, these critics assumed that the lack of patterns shown by trees in the humid East represented the normal state of affairs and that Douglass's records were in some way faulty. Attacked from two directions, Douglass agreed with Carnegie official Frederic E. Clements that the meeting probably did more harm than good. He later recorded his disappointment by writing that "Many spoke without knowing what they were speaking against" and "My assertions ridiculed tho based on 20 years of work under special conditions. They were ignorant and didn't know it."[24]

Although unable to establish his cycle studies within the scientific mainstream, Douglass remained convinced of the value of his research. The Carnegie Institution, who had supported the Tucson astronomer in his tree ring efforts since 1918, stood by him. During the summer and fall of 1934, Carnegie President John C. Merriam successfully lobbied the trustees of the institution for a major program to allow Douglass to devote full time to his research. Announced by University of Arizona President Homer L. Shantz in early February 1935, the three-year program would provide Douglass with sufficient funding for his own salary, that of two assistants and clerical staff, and equipment. Douglass was to focus on cycle analysis, with the goal of long-range weather forecasting. The University of Arizona would continue to retain Douglass as a member of its faculty but the university's primary contribution would be to provide laboratory and office space on campus.[25]

Among the immediate goals of the new program was the preparation of a third volume in the Carnegie series on climatic cycles. During the summer of 1935, Douglass drafted an outline for the book (subtitled "A Study of Cycles") and soon began writing. He completed the first draft of the volume at the end of November and sent a revised version to Merriam in early January 1936. By the end of the month, Carnegie officials had authorized funds for printing a thousand copies of Douglass's book, but revisions proved far more involved than anticipated. Douglass mailed copies of his manuscript to various colleagues in astronomy and other disciplines while reworking his discussion of the data and analysis that underlay his research. Corrections and revisions were sent from Tucson to Carnegie headquarters in Washington throughout the spring and summer, with proof corrections finally completed in November. Douglass received his bound copy of *Climatic Cycles and Tree Growth*, volume 3: *A Study of Cycles* a few days before Christmas 1936.[26]

The completed third volume of *Climatic Cycles and Tree Growth* presented the results of more than three decades of research, but clearly focused on cyclical phenomena and their potential value to long-range weather forecasting. This focus was indicated in the foreword, written by Harvard statistician Edwin B. Wilson. At the National Academy of Sciences conference in 1932, Wilson had declared that Douglass's analytical method was of potential value and should be taken seriously. He repeated this analysis in his foreword, later explaining to Douglass that "my main aim was to impress upon various people that they had not given sufficient attention to your work." Douglass introduced his work by emphasizing that the purpose of the volume was to describe and discuss the new method of cycle analysis, as well as to show the results of tree ring research and to begin the study of climatic cycles in long-range weather prediction. As he had done for several years, Douglass stressed the value of cycles for forecasting, even if they could not be described by mathematical equations. "It is here believed," Douglass wrote, "that it is a mistake to neglect the examination of climatic changes because they do not seem permanent and exactly timed and are not molded in an artificial form like a sine curve."[27]

Douglass presented the results of his work in seven chapters, all but the first of which stressed the analytical efforts which had guided his research. He described the cyclogram method and showed how such analysis revealed cyclical patterns in solar and terrestrial records, as well as the relationships between the two. His final chapter, "The Cycle Problem and Long-Range Forecasting," repeated his emphasis on the value of non-permanent cycles and their analysis for climatic study. Especially from the study of Arizona data, it was clear to Douglass that climatic variations represented a "cycle complex" that would need to be studied further to be completely understood. Various aspects of the tree ring record, even in the extensive data from Arizona pines, remained unclear. Thus, prediction remained uncertain. Given the more meager data from the Mississippi Valley and the Pacific Coast, prediction for those regions was even more problematical. Nonetheless, the cyclical patterns continued to strike Douglass as an important datum in his effort to establish a base for long-range weather forecasting. Particularly intriguing were the similarities between climatic and solar cycles, which strongly suggested to Douglass a connection between radiation and terrestrial climate. This connection too, Douglass insisted, required more data and further study.[28]

The appearance of the third volume of *Climatic Cycles* paralleled the beginning of the final year of the Carnegie program at the University of Arizona and of a new status for Douglass. Facing his seventieth birthday in July, he informed University of Arizona President Paul S. Burgess in early 1937 of his desire to assume emeritus status. Douglass also emphasized that the university needed to assume the responsibility for dendrochronological research following the end of the Carnegie program. A tentative move in this direction took place during the spring when Douglass was provided rooms and storage space in the new football stadium on campus. For the first time, his tree ring research was concentrated in one building. As the year approached its end, however, Douglass forwarded various plans and proposals to university officials for a permanent dendrochronology facility to continue the work which would soon lose its Carnegie support. The university and the Arizona Board of Regents recognized the value of Douglass's efforts. The board established the Laboratory of Tree-Ring Research in early December and appointed Douglass as director. The university budget allowed only minimal support for the new facility, however, and Douglass remained unsuccessful in his efforts to obtain outside funding. Although he persuaded Carnegie officials to leave all equipment in Tucson on a permanent loan, the new laboratory remained in a precarious financial position for many years.[29]

Despite Douglass's inability to convince certain colleagues of the value of his cycle concept and his new method of analysis, the possibility of climatic prediction proved of great interest. Throughout the 1930s, Douglass received requests for information concerning rainfall patterns. Officials with the Tennessee Valley Authority and other New Deal agencies asked for any data and insight Douglass might have, as did engineers seeking to improve water supplies in Denver and the Colorado River watershed.[30] The establishment of the Laboratory of Tree-Ring Research led to increased requests and various contracts for research programs. In 1940, for example, the U.S. Department of the Interior financed a tree ring collecting trip to the upper Gila River area to determine rainfall patterns of use in the management of Coolidge Dam.[31] Even more dramatic was the laboratory's involvement with the Los Angeles Bureau of Power and Light during World War II. To more efficiently manage Boulder Dam and its production of electrical power for southern California war industries, the bureau negotiated a contract with the laboratory in September 1941 for the collection and analysis of tree ring specimens from the Colorado River drainage area. Edmund Schulman, who had worked with Douglass for nearly a decade, traveled extensively in the area during the fall and spent the next six months carefully analyzing the extensive tree ring collection. Schulman's report, submitted in May 1942, concluded that the region suffered approximately five severe drought years per century and that the average length of a period of notable excess or deficiency in rainfall was on the order of nine years. Schulman stressed that such predictions were only statistical probabilities, but his work at least provided Boulder Dam management with a guide for regulating the amount of water flowing through the facility.[32]

Although Douglass's goal of precise long-range weather forecasting remained unfulfilled, dendrochronology represented an important addition to the scientific study of climate. Much of the research conducted by the Laboratory of Tree-Ring Research has focused on such studies and has provided valuable predictive information.[33] Douglass was less successful in his efforts to convince scientific colleagues of the legitimacy of his cycle concept and analysis techniques. The nonmathematical description of cycles, the existence of several cycles of different values, and the use of unorthodox equipment left critics with doubts about Douglass's research and findings. Nonetheless, the science of dendrochronology and the institutional base Douglass established for it contributed an important new perspective for the study of terrestrial climate.

Part II The Scientific Community

Introduction

The first half of the twentieth century defined the modern Southwest. Political, economic, and demographic changes transformed the region from a sparsely populated colony symbolized by its territorial status to an increasingly visible participant in America's postwar expansion. Although "Sun Belt" had not yet emerged as the appellation often used to describe the Southwest's new status, the characteristics associated with this phrase had become well established by mid-century. For the next five decades, the region would play a national role built upon this foundation.

The Southwest's population growth between 1900 and 1950 represented the region's most dramatic characteristic. Defining the region as Arizona, New Mexico, and the western panhandle of Texas, the Southwest's population increased from approximately 350,000 in 1900 to nearly 1.5 million in 1950. Although the region remained one of the least populated areas of the United States, its 1 percent share of the nation's population in 1950 was more than twice its share fifty years earlier. Economic changes were also dramatic. As the twentieth century opened, the region was primarily devoted to extractive industries such as mining and agriculture. Although these pursuits remained central to the region's economy at mid-century, they were augmented by the rapidly growing electronics industry, expanding federal installations, and the burgeoning service industries made necessary by population growth. Similarly, the region's political significance grew throughout the period, with individuals such as Carl Hayden, Bronson Cutting, and Clinton Anderson playing an increasingly important role in national politics.[1]

Although rarely acknowledged, the pursuit of science in the American Southwest was also an important part of the region's maturation. In 1900, the Southwest remained little more than a source of scientific data. Fifty years later, the Southwest could boast of dozens of important scientific installations, including federal complexes such as the Los Alamos Scientific Laboratory, private facilities such as the Lowell Observatory, and the state universities in Tucson, Albuquerque, Las Cruces, and El Paso. Virtually every scientific discipline was represented in these institutions, providing the Southwest with a secure foundation for the dramatic growth that characterized the following decades.

As the twentieth century began, science in the American Southwest remained in a nebulous state. The previous decades had witnessed significant interest in the region as a source of new species of plants and animals. Expeditions financed by the federal government brought specimens from the region to eastern museums for study and preservation. The human inhabitants of the region were also the focus of much scientific study. Archaeology and anthropology became increasingly important activities in the Southwest during the last two decades of the nineteenth century. Government agencies such as the Smithsonian Institution and the U.S. Geological Survey stimulated the study of existing and prehistoric peoples, as did various privately funded expeditions. Harvard's Peabody Museum sent several parties to explore the region, as did the Archaeological Institute of America. Wealthy patrons financed such programs as the Hemenway South-Western Archaeological Expedition and the Hyde Exploring Expeditions, which surveyed many of the sites later central to archaeological research. These efforts included many of the leading figures of the early years of southwestern archaeology. John Wesley Powell, Adolph Bandelier, Frank Hamilton Cushing, Richard Wetherill, Jesse Walter Fewkes, and Edgar Lee Hewett were all involved in these pioneering studies.[2]

The surveys of the region represented a colonial perspective on science; eastern institutions supplied funding and personnel and claimed the results of these investigations. Late in the nineteenth century, however, tentative efforts to establish a regional base for science emerged. These efforts focused on institutional developments, increasingly necessary for the advance of science. Although the Lowell Observatory in Flagstaff, Arizona, represented a research outpost in the region, educational institutions emerged as the foundation for science in the Southwest. During the late 1800s, Arizona and New Mexico both attempted to establish systems of higher education to provide an intellectual base and to show that the region was worthy of ultimate statehood. In both territories, however, the path to their educational goals proved a difficult one.

In 1889, the territorial legislature of New Mexico established three institutions, all of which were provided with land grants for financial support. Albuquerque business leaders successfully lobbied for the territorial university, while Las Cruces received the agricultural college and Socorro the school of mines. The agricultural college opened in 1890 with seventeen students, followed two years later by the other institutions. The University of New Mexico began with only two divisions, a normal department to train teachers and a preparatory program to provide high school instruction. Although the practical focus of the Las Cruces and Socorro schools gave them a clear mission, the Albuquerque facility struggled during its early years. Enrollments remained low in all these schools, with student populations measured in the dozens rather than in the hundreds.[3]

The situation in Arizona was even less encouraging. The legislature established a university and a normal school in 1885, but by the end of the decade only the normal school in Tempe had begun to offer regular instruction. In Tucson, the University of Arizona was still struggling to organize itself in 1889 and had not even completed its main building. By the end of the nineteenth century, both Arizona schools were functioning at acceptable levels and had been joined by Northern Arizona Normal School in Flagstaff. As was the case in New Mexico, enrollments remained low and the schools supplied few opportunities for the pursuit of scholarship by faculty. Heavy teaching loads and the lack of suitable equipment were particularly troublesome to the scientists who were hired by these territorial institutions.[4]

From the perspective of southwestern science in the late nineteenth century, the significance of educational institutions came not through the universities themselves but through the agricultural experiment stations that were associated with the colleges in Las Cruces and Tucson. Federal funding was the key to these facilities, which supplied employment and support to scientists investigating topics of value to the territorial economy. Indeed, it is impossible to separate the experiment stations from the universities that housed them. Of the $42,000 appropriation for New Mexico College for 1890–91, $30,000 was in federal funds from the Morrill Act (land grant institutions) and the Hatch Act (experiment stations). Hiram Hadley, first president of the college, was also director of the experiment station during his 1889–1894 tenure. The next four leaders of the Las Cruces school followed this precedent, clearly indicating the importance of the station to the college. In Tucson, the availability of Hatch Act funds was seen as an opportunity to complete the territorial university. The Board of Regents established a College of Agriculture (without a building, faculty, or curriculum) and applied for Hatch Act funds to establish an experiment station under it. The $10,000 that arrived in June 1890 not only established the station, but also provided the university with faculty and operating expenses.[5]

Arizona's creative application of Hatch Act funds was not unique in the late nineteenth century. When the Hatch Act was passed in 1887, it represented a clear attempt to apply science to American agriculture to make farmers more competitive. By providing federal funding to support agricultural experiment stations, Congress hoped to encourage applied scientific research and the dissemination of results. The stations, however, were rarely able to pursue the stated goals of the program. Because they were part of generally underfunded land-grant colleges, experiment stations often found their budgets pirated for other purposes. Basic research was largely ignored, as local residents and politicians viewed the experiment stations as "model farms" to perform experiments and run tests that farmers could not. Scientists attached to the stations thus found themselves analyzing soil samples sent in by local growers and teaching courses in botany, zoology, and chemistry in the regular university curriculum. Nonetheless, the stations did offer employment for scientists, who increasingly found their research focus applauded by the federal officials in charge of scientific programs in the Department of Agriculture.[6] In the Southwest, the experiment stations played an especially important role in establishing a scientific presence in the region.

Although the institutional base is crucial to an awareness of scientific developments in the Southwest, scientists themselves established the place of science in the region. An analysis of the Southwest's scientific community is thus essential to an understanding of developments in the twentieth century. The raw material for such an analysis is available in the various volumes of *American Men of Science* published between 1906 and 1949, which offer a valuable portrait of the American scientific community during the first half of the twentieth century.

These volumes were designed by James McKeen Cattell (1860–1944), professor of psychology at Columbia University and editor of the journal *Science*, as a biographical directory of the American scientific community. He began collecting data on North American scientists in 1903, distributing some ten thousand requests for information to individuals identified through membership lists of professional organizations and notices in such periodicals as *Science, Popular Science Monthly*, and *The Nation*. Cattell's goal was to identify the leading American scientists, focusing on those individuals "who have carried on research work in the natural and exact sciences." The 4,000 listings in the first volume of *American Men of Science* grew steadily over the next few decades, reflecting the growth of the American scientific community and the greater recognition of the directory's value by scientists. Entries grew to 9,500 by 1921 and reached 28,000 by the 1938 edition. More than 50,000 appeared in the 1949 volume, indicating the significant growth of the scientific community during the 1940s. Cattell also identified the leading research scientists in various disciplines, relying on the evaluation of specialists in these fields. The 1,000 individuals thus identified were marked by stars in the volumes, with later additions nominated by those previously starred.[7]

As is the case with any biographical directory, *American Men of Science* may be criticized for various shortcomings. Cattell's selection criteria eliminated various disciplines from serious consideration (relatively few physicians and engineers were included) and arguably instituted an eastern bias into the listings. Because individuals supplied their own biographical information, omissions and inaccuracies occasionally emerged. The starring procedure remained largely unchanged during Cattell's editorship (it was discontinued in the 1949 edition), ignoring changes in disciplinary distribution during the first half of the twentieth century. Nonetheless, as pointed out by various historians, these volumes provide an excellent overview of the American scientific community.[8] When used carefully they offer a valuable guide to the growth of science during the period.

Those scientists living in the Southwest and listed in these volumes thus represent a subset of the nationally recognized scientific community and serve as a useful base for an analysis of the region's science. The steady growth of southwestern listings indicates that the region was establishing itself in science throughout this period. From 15 in the 1906 edition, the number of scientists grew to 26 (1910), 52 (1921), 71 (1927), 169 (1933), and 214 (1938). In 1949, largely as a result of the fundamental changes in the region's science brought about by World War II, southwestern scientists numbered 502. When combined with a discussion of contemporary institutional changes, an analysis of the region's scientists as described in *American Men of Science* provides an intriguing portrait of a scientific community undergoing dramatic growth and maturation. By mid-century, this community had established itself as an important factor in American scientific pre-eminence.

Chapter 4 Foundations and Institutions, 1900–1940

At the beginning of the twentieth century, science in the Southwest displayed intriguing characteristics. As had been the case during the nineteenth century, botanists, zoologists, and archaeologists viewed the region as a natural laboratory. Previously unknown species of plants and animals were observed and cataloged, adding to scientists' knowledge of North American life forms. Evidence of ancient civilizations attracted the interest of archaeologists and anthropologists, who were able to explore the well-preserved ruins throughout the region and to investigate contemporary Native American cultures. With few exceptions the scientists exploring the Southwest were visitors, investigating the region as representatives of eastern institutions. Data from such scientific visits, whether specimens of new species or artifacts from pueblo ruins, were dispatched to the institutions supplying the personnel for the scientific investigation of the American Southwest.

Over the next four decades, however, the scientific landscape of the American Southwest underwent important changes. Institutional growth provided scientists with employment and a visible if limited base of support for their work. The region's universities grew steadily during this period, and were augmented by a wide variety of additional facilities. Private bequests supported such institutions as the Lowell and McDonald observatories, the Museum of Northern Arizona, and the Laboratory of Anthropology in Santa Fe. The federal government became an increasingly visible factor in the region's scientific growth, providing support through programs of the Agriculture and Interior departments. As a result of these institutional factors, the region's scientific community grew dramatically. In the first edition of *American Men of Science*, published in 1906, only 15 of the more than 4,000 entries lived in the Southwest. The 1938 edition of this biographical directory, on the other hand, included more than 200 scientists from the Southwest among its 28,000 entries. On the eve of World War II, the Southwest had laid the foundations for science in the region and had established a base for later dramatic growth.

Laying Foundations: 1900–1920

Although the first two decades of the new century witnessed various changes in the Southwest, science remained a following, rather than a leading phenomenon. Signs of potential emerged, however, especially in terms of institutional growth. Universities in the region continued to represent the region's intellectual focus, despite low enrollments and limited resources. Although the University of Arizona remained the largest institution of higher education in the region, its student population did not approach one thousand until the end of this period. Other institutions grew more slowly, generally with enrollments of less than three hundred. New Mexico School of Mines rarely enrolled more than a hundred students, but assumed an important role in the development of the region's science when it began the Mineral Resources Survey of the state in 1915. The major change in the region's higher education was the establishment of the Texas State School of Mines and Metallurgy in El Paso. Authorized by the state legislature in 1913, the school opened in the fall of 1914 after El Paso citizens donated $50,000 to purchase land for the institution. Growing slowly over the next several years, the El Paso school focused on the needs of the mining industry and provided employment for scientists and engineers.[1]

The agricultural experiment stations associated with the universities in Tucson and Las Cruces remained crucial for science in the Southwest. Reflecting the practical focus of the region's higher education, the stations hired scientists to pursue activities designed to aid agriculture. Dry farming techniques, irrigation, and the cultivation of such crops as dates, cotton, olives, chiles, wheat, and beans were important topics pursued by experiment station staff members. The stations hired chemists, entomologists, engineers, and botanists to improve the region's agriculture through scientific research and applications, as well as to transmit the results of their studies to southwestern farmers through publications, farmers' institutes, and demonstration trains. These scientists' activities paralleled the increasing emphasis on research that characterized experiment stations elsewhere in the United States. Passage of the Adams Act in 1906 capped more than a decade of efforts by leading experiment station scientists to redirect the stations' focus to original research. Taking advantage of a growing supply of academic scientists from leading graduate programs, the stations could now pursue research that had potential value to agriculture rather than research designed to solve specific and immediate agricultural problems.[2]

Other examples of institutional growth appeared in the Southwest during the early twentieth century. The region's value as a source of unique plant specimens led to the establishment of the Desert Botanical Laboratory, funded by the Carnegie Institution of Washington and constructed in the desert west of Tucson. The choice of Tucson for this laboratory in 1902 reflected more than geography, however. The presence of the University of Arizona and Tucson's transportation advantages were both cited as factors in the choice. Guided by noted botanist Daniel T. MacDougal, the laboratory soon included such figures as botanist Forrest Shreve (Ph.D., Johns Hopkins University) and plant physiologist Herman A. Spoehr, both of whom became noted figures in the development of American botany during the next few decades.[3]

The federal government increased its institutional presence in southwestern science through activities of the Department of Agriculture. The Bureau of Plant Industry established its Cotton Research Center at Sacaton, Arizona, in 1907 and began research that ultimately led to the perfection of pima cotton. The Forest Service established its first forest experiment station the following year on the Coconino Plateau near Flagstaff. Its efforts to apply scientific knowledge to the various problems of American forests proved very successful and led to the establishment of similar stations in Colorado, Idaho, Washington, California, and Utah. The Forest

The Carnegie Desert Botanical Laboratory as it appeared in 1911. (Courtesy of the Arizona Historical Society/Tucson; AHS #43988)

Service provided scientists with various employment opportunities. Between 1909 and 1924, for example, Yale Forestry School graduate Aldo Leopold held several positions in Arizona and New Mexico. His efforts in range and wildlife management played an important role in the development of Leopold's conservationist perspective, as expressed in published works during the next few decades.[4]

Growing interest in southwestern archaeology also played an important role in better establishing science in the region. Largely through the work of Edgar Lee Hewett, the early twentieth century witnessed the beginnings of an institutional base for the region's archaeological studies. A self-trained archaeologist (although he later earned a doctorate from the University of Geneva), Hewett was instrumental in forming the New Mexico Archaeological Society in 1900 and coordinated efforts which led to the establishment of the School of American Archaeology (later renamed the School of American Research) and the Museum of New Mexico. In Tucson, the University of Arizona established the region's first anthropology department in 1915, augmenting the Arizona State Museum and

Table 4.1 Principal Employers of Southwestern Scientists, 1900–1940

Institution	American Men of Science Volume					
	1906	1910	1921	1927	1933	1938
University of Arizona	5	11	14	28	59	61
Lowell Observatory		1	5	3	3	5
Carnegie Desert Laboratory	1	3	8	5	3	3
University of New Mexico	2	3	3	6	11	23
New Mexico College	1	1	1	2	17	16
New Mexico School of Mines	1	2		2	5	5
Texas School of Mines				1	5	4
U.S. Department of Agriculture			5	8	15	23
U.S. Department of the Interior			3	3	6	9

providing an educational focus for the study of the region's ancient inhabitants.[5]

Despite the nebulous status of science in the Southwest in the early twentieth century, the region could claim a nucleus of scientists. As indicated by the *American Men of Science* volumes published in 1906 and 1910, this nucleus was a small one, but an analysis of these scientists provides a useful portrait of the nascent scientific community. Nearly half the scientists listed in these volumes lived in Tucson, reflecting the strong institutional base of the city. The University of Arizona, the associated experiment station, and the Carnegie Desert Laboratory employed scientists trained in various disciplines. In New Mexico, only the territorial university in Albuquerque and the School of Mines in Socorro employed more than one scientist among those listed, while west Texas could claim only two entries in 1906 (table 4.1). Southwestern scientists represented a disciplinary concentration as well (table 4.2). More than half pursued work in the fields of agriculture, botany, geology, or chemistry, all of which had significant practical applications in the region as well as academic value. Among the distinguished scientists who were "starred" in these volumes, botany and geology were the prevalent disciplinary affiliations (table 4.3). Educational characteristics indicate that the region's scientists were not

Table 4.2 Specialties of Southwestern Scientists, 1900–1940

	American Men of Science Volume					
Discipline	1906	1910	1921	1927	1933	1938
Agriculture	2	1	1	7	26	30
Anthropology and archaeology		1	2	3	3	15
Astronomy	1	2	7	6	6	11
Biology	6	8	18	17	34	43
Chemistry		5	4	7	15	16
Engineering		1	3	4	16	15
Forestry			2	2	5	8
Geology	3	4	5	9	19	21
Mathematics	1	1	2	4	11	19
Medicine			2	1	12	11
Mining and metallurgy	1		2	3	6	8
Physics		1	1	4	6	8
Psychology		1	2	3	4	8
Other	1	1	1	1	6	2
Total	15	26	52	71	169	215

educated very differently from their colleagues elsewhere in the United States. All but a few of the southwestern scientists held undergraduate degrees, more than half had earned at least one postgraduate degree, and slightly less than half held doctoral degrees. Southwestern scientists of this period had attended a wide variety of undergraduate institutions, including such leading schools as Yale and the universities of Chicago, Michigan, and Wisconsin (table 4.4). Columbia, Johns Hopkins, Chicago, and Clark supplied the Southwest's scientists with Ph.D. or similar degrees (table 4.5). These four institutions were among the leaders in the production of science doctorates at the national level as well.[6]

The handful of starred scientists listed in these two volumes reflects the colonial characteristics of the region's science. Although currently living in the Southwest, these individuals had generally earned distinc-

Table 4.3 Specialties of Southwestern Scientists Starred in *American Men of Science*, 1900–1940

Discipline	*American Men of Science* Volume					
	1906	1910	1921	1927	1933	1938
Astronomy			3	4	5	5
Botany	1	2	4	2	3	1
Geology	2	1			2	3
Mathematics	1					1
Medicine					1	
Psychology		1				
Total	4	3	7	6	11	10

tion through work performed elsewhere. William P. Blake, for example, served as one of the West's leading consulting geologists throughout the late nineteenth century before joining the faculty of the University of Arizona in 1895. Eighty years old when he was listed in the first edition of *American Men of Science*, Blake was essentially retired at this point. Botanist William A. Cannon was one of the few scientists starred for work in the Southwest. A doctoral graduate of Columbia University, Cannon began his association with the Carnegie Desert Laboratory in 1903, becoming a staff member two years later. His research concerning the root systems of desert plants led to his election as a fellow of the American Association for the Advancement of Science and the award of a star in the 1910 edition of *American Men of Science*. For the most part, nonetheless, scientists in the American Southwest had not yet had opportunities to make the scientific contributions that would lead to national recognition.

The American Southwest did offer opportunities for diverse scientists who might not otherwise have risen to professional prominence. An excellent example of an individual who took advantage of such opportunities was Fabián García, whose career at the New Mexico Agricultural Experiment Station lasted nearly half a century.

Table 4.4 Principal Undergraduate Programs of Southwestern Scientists, 1900–1940

Institution	*American Men of Science* Volume					
	1906	1910	1921	1927	1933	1938
University of Wisconsin	1	1	2	3	3	5
University of Illinois	1	2	1	1	5	4
University of Michigan	1	1	2	1	3	6
University of Chicago		1	2	5	5	7
Indiana University		1	4	4	5	5
University of Minnesota		1	1	2	3	5
University of California			1	1	4	7
University of Nebraska			2	3	2	4
Kansas State College				1	4	5
University of Arizona					8	17

Born in Chihuahua, Mexico, in 1871, García was brought to the United States by his grandmother two years later, shortly after the death of his parents. He was a member of the first class of New Mexico College of Agriculture and Mechanic Arts in Las Cruces, earning his B.S. degree in 1894. Following graduation, García joined the experiment station staff and spent the 1899–1900 academic year at Cornell University in a program of independent study. He completed requirements for the Master of Science in Agriculture from New Mexico College in 1905 and was appointed professor of horticulture the following year. García also served as horticulturist for the experiment station from 1906 until 1913, when he became director of the facility. For the next thirty years, García coordinated the activity of the experiment station, gave numerous lectures as part of the station's extension work, and pursued various research projects designed to expand and improve New Mexico agriculture.[7]

Although his early work focused on orchard crops such as plums, peaches, apricots, and cherries, García's most noteworthy research program began when he initiated chile breeding experiments in 1907. Existing varieties suffered from inconsistent pod size and pungency, retarding

Table 4.5 Principal Doctoral Programs of Southwestern Scientists, 1900–1940

Institution	American Men of Science Volume					
	1906	1910	1921	1927	1933	1938
University of Chicago	1	1	2	5	11	18
Columbia University	2	2	2	1	2	3
Johns Hopkins University	1	1	1	1	3	5
University of Wisconsin		1	1	3	8	12
University of California			1	4	7	10
Stanford University			2	3	2	4
University of Illinois			3	1	3	3
University of Minnesota			1	3	3	3
Harvard University					3	7

the development of a viable chile industry. His initial research led to the publication of *Chile Culture* in 1908, the first of many experiment station bulletins on the topic. During the next decade, García pursued a breeding program using three types of chile: a red chile brought from California in 1902, a black variety imported from Mexico in 1903, and pasilla chile from Chihuahua. By 1913, he had selected two of the many crossed varieties as most desirable, ultimately choosing the chile that would be called "New Mexico No. 9" three years later. Further development yielded a milder chile (to expand the market beyond the region) with a pod shape more suitable for peeling and canning. An unexpected benefit from García's breeding program was that this pepper proved more resistant to chile wilt than local varieties. Released in 1921, "New Mexico No. 9" established the state's chile industry and served as the standard cultivar until the 1950s.[8]

García continued to work with chiles and expanded his research to include the cultivation of grapes, beans, pecans, and pears. He reported the results of investigations concerning the impact of climatic change on crops and the value of smudging to protect orchards from cold.[9] García's investigations provided area farmers with valuable information on a wide variety of agricultural topics. Under his directorship, the New Mexico Agricultural Experiment Station fulfilled its function as an institution designed to disseminate the results of scientific research among farmers and other agricultural interests.

Building Communities: The 1920s

The third decade of the twentieth century saw the institutional base for the region's science grow steadily, reflecting the general growth enjoyed by the area. The number of southwestern scientists increased as well, with seventy-one individuals listed in the 1927 edition of *American Men of Science*. Not only were there more scientists, they increasingly represented a wide variety of disciplinary interests and backgrounds. More significant still, scientists living in the American Southwest reached a critical mass and emerged as a community during this decade. While establishing themselves within the national scientific community, these scientists began to define themselves in regional terms as well. This self-definition in part compensated for the isolation of the sparsely populated Southwest and further established a base for the growth of the region's scientific community.

The 1920s witnessed noteworthy cultural growth throughout the West, paralleling dramatic population gains. Although higher education in Arizona, New Mexico, and west Texas rarely received generous funding, the universities in these states nonetheless established themselves on a firmer base during the period. The University of Arizona continued to be the region's leading educational institution and expanded more rapidly than other schools in the Southwest. Enrollment more than doubled, exceeding 2,000 students by the end of the decade. Faculty increased from fewer than 100 to more than 150 by the late 1920s, while library holdings doubled to some 70,000 volumes. New Mexico schools grew less dramatically. The University of New Mexico enrolled some 700 students by the late 1920s, more than twice the number at the agricultural college in Las Cruces. The School of Mines in Socorro maintained a student population of fewer than a hundred during the decade, but gained an important boost for its science program when the State Bureau of Mines and Mineral Resources became part of the school complex in 1927. The Texas College of Mines and Metallurgy passed the 500 mark in enrollment the same year, offering an undergraduate degree in mining engineering and adding liberal arts courses to the curriculum.[10]

Other institutions in the Southwest remained central to the advance of science. The Carnegie Desert Laboratory hosted visiting scientists, most of whom used the facility as a base for their field and laboratory research concerning desert botany and ecology. Desert Lab staff member Forrest Shreve attracted significant national attention for his work, as indicated by his 1922 election as president of the Ecological Society of America. In Flagstaff, the Lowell Observatory was building its reputation among American astronomers. No fewer than five staff members appeared in the 1921 *American Men of Science* volume, two of whom had been starred for their work at the observatory. Federal agencies were also more evident in the region during the 1920s. The increasing importance of scientific training to the work of government programs provided employment opportunities for southwestern scientists with the Department of Agriculture, the Bureau of Mines, and the U.S. Geological Survey.[11]

During the 1920s the growing population of scientists organized both locally and regionally to establish a sense of community and a forum for the exchange of ideas. In 1923, for example, naturalists in southern Arizona formed the Tucson Natural History Society. Attracting university faculty as well as amateur biologists, the group met monthly, sponsored excursions to local areas of interest, and scheduled lectures by local and visiting biologists. Very interested in conservation, the society soon persuaded the U.S. Forest Service to establish a natural area in the Catalina Mountains north of Tucson.[12]

The most dramatic example of the efforts to establish a sense of community in the Southwest was the organization of the Southwestern Division of the American Association for the Advancement of Science. These efforts began in January 1919, when physician E. C. Prentiss read a paper before the El Paso County Medical Society calling for a regional scientific association. Meetings and correspondence followed during the next several months, involving many of the region's scientists and ultimately leading to an organization meeting in April 1920 on the University of Arizona campus. Although only nine individuals attended this meeting, they represented a cross-section of the region's scientists. Daniel T. MacDougal, for example, not only attended as a member of the Carnegie Desert Laboratory, but also served as a delegate from the Executive Committee of the national association. University of Arizona astronomer A. E. Douglass served as secretary of the meeting which Prentiss chaired. Scientists without academic affiliations were also on hand, as shown by Phoenix consulting engineer Arthur I. Flagg. He would serve in various positions during the 1920s and was elected president of the division in 1927.

With guidance from MacDougal, the meeting drafted a constitution and defined the division as Arizona, New Mexico, Texas west of the Pecos River, and the adjacent states of northern Mexico. The meeting elected officers to serve until the first annual meeting, selecting Edgar Lee Hewett of Santa Fe as president, Elliott C. Prentiss as vice president, and A. E. Douglass as secretary-treasurer. Douglass was charged with the task of writing to other scholarly groups in the region to determine if they would be interested in becoming affiliated societies. He was also directed to write to the American Association for the Advancement of Science to urge them to recommend the establishment of a national monument west of Tucson to protect the large stand of saguaro cacti in the area. By the first annual meeting in El Paso in early December 1920, the Southwestern Division had attracted the attention of many of the region's scientists (143 individuals were listed as members) and had organized a program to reflect the wide variety of interests represented by the diverse membership.[13]

Over the next decade, the Southwestern Division served as an important institutional focus for the region's scientists. Meeting in such cities as Tucson, Santa Fe, Phoenix, Albuquerque, and Boulder, Colorado (after Colorado was added to the division), the organization attempted to unite the region's scientists. Presidents of the Southwestern Division included many of the leading scientists of the region, such as A. E. Douglass, Vesto M. Slipher, and Forrest Shreve. Although membership failed to grow significantly during the decade, the Southwestern Division played an important role in establishing science as a noted part of the region's growth and cultural maturation.[14]

The southwestern scientists listed in the two editions of *American Men of Science* published in the 1920s reflected the changing face of the region's scientific community. Regional listings grew to fifty-two in 1921 and seventy-one six years later, but displayed characteristics suggesting a more diverse and more mature community. The University of Arizona and the University of New Mexico remained the dominant employers of scientists among academic institutions, but a large number of scientists were also affiliated with such facilities as the Carnegie Desert Laboratory and the Lowell Observatory. The federal government employed several scientists throughout the region, primarily through the Department of Agriculture (table 4.1). The educational backgrounds of the region's scientific community remained little different from those of their colleagues elsewhere. At the undergraduate level, the state universities of Indiana, Michigan, Minnesota, Nebraska, and Wisconsin dominated the list, although private schools such as the University of Chicago and Yale University also produced several southwestern scientists (table 4.4). At the doctoral level, southwestern scientists came from the leading programs at Chicago, Columbia, Wisconsin, and a few others, although the increasingly well-known programs at Stanford and Berkeley also were important (table 4.5). The doctoral origins of southwestern scientists were similar to those of their colleagues elsewhere in the United States.[15]

The region's scientists showed even more clearly than before a disciplinary concentration. The leading disciplines were in such areas as agriculture, biology, geology, and chemistry, all of which offered opportunities for academic contributions and applied research. An entomologist, for example, could find employment as a faculty member in the biology program of a state university, as a staff member of an experiment station, or as a research specialist for the U.S. Department of Agriculture. Geologists found employment in the region's mining schools (Tucson, Socorro, El Paso) and with the numerous mining companies in the Southwest, as did engineers and metallurgists. Astronomers continued to be a significant part of the region's scientific community, taking advantage of the favorable climate of the Southwest (table 4.2). Among starred scientists, a similar concentration was evident. These distinguished members of the region's scientific community were all either astronomers or botanists, indicating the continuing importance of the natural characteristics of the region for scientists' significant contributions (table 4.3).

Starred scientists listed in these two editions of *American Men of Science* displayed an important change in the region's scientific status. Previously, the region's distinguished scientists had earned notoriety for research and other activity performed before they arrived in the Southwest. This remained somewhat the case in the 1920s, as indicated by such figures as botanists Daniel T. MacDougal and Frederic E. Clements, both of whom had been starred before joining the Carnegie Desert Laboratory in Tucson. Three of the southwestern scientists starred in the 1921 edition of *American Men of Science*, however, had been recognized for their work in the region. Botanist Forrest Shreve gained notoriety for his work on desert plants, while astronomers Vesto M. Slipher and Carl O. Lampland were recognized for their numerous contributions to astronomy from the Lowell Observatory. Investigating planetary, stellar, and nebular phenomena, Slipher and Lampland had attracted international attention to the research programs of the Flagstaff facility. These starred scientists were joined in the 1927 volume by astronomer A. E. Douglass of the University of Arizona. Douglass was recognized for his efforts as director of the recently completed Steward Observatory, but also for his early work in establishing the new science of dendrochronology.[16]

Establishing Institutions: The 1930s

Although the Southwest endured the same Depression-era difficulties as the rest of the United States, the 1930s proved to be an auspicious decade for science in the region. Not only did the scientific community (as reflected in the 1933 and 1938 editions of *American Men of Science*) grow dramatically, but also a similar growth was evident in the institutions supporting science in the Southwest. Existing facilities took advantage of federal funding opportunities, while new institutions emerged through federal, state, and private financial support. Educational institutions benefited from New Deal policies and growing enrollment, securing their place as a key part of the foundation for southwestern science. The University of Arizona utilized a faculty of nearly 300 by the end of the decade to educate more than 3,000 students; the University of New Mexico was only slightly smaller. Other schools in the region claimed much smaller enrollments, but also showed impressive advances. New Mexico College in Las Cruces grew to more than a thousand students, while the School of Mines in Socorro expanded its physical plant through the programs of such New Deal agencies as the Works Progress Administration and the Public Works Administration. The Texas School of Mines became a four-year college in the fall of 1931 and immediately hired five new faculty members with doctorates to upgrade the curriculum. This school also exceeded one thousand by the late 1930s.[17]

Although the region's universities remained important, the growth of other institutions during the 1930s contributed even more to the advance of science in the region. Arguably the most dramatic development during this period was the completion of the McDonald Observatory on Mount Locke, Texas. This facility emerged in the early 1930s as a joint effort of the University of Texas (to whom banker William J. McDonald had willed nearly one million dollars) and the University of Chicago. The latter's Yerkes Observatory was one of the world's leading astronomy centers, but its Wisconsin location compromised many important observations. The superior astronomical climate in west Texas would provide Yerkes astronomers with significant advantages, while their expertise in designing and maintaining telescopes would provide the University of Texas with many benefits. Construction on Mount Locke began at the end of 1933, followed a year later by the erection of the first telescope on the mountain, a twelve-inch refractor that was immediately put to use for observations of variable stars. The resident staff on the mountain grew slowly over the next few years, but several Yerkes astronomers observed from Mount Locke. Gerard P. Kuiper, for example, who would later become one of the nation's leading astronomers, divided his time during the 1936–37 academic year between Yerkes and McDonald to pursue his work on double stars. The completion of the eighty-two-inch McDonald reflector during the spring of 1939 secured a leading role for the west Texas facility in astronomical research.[18]

Institutional developments in Tucson followed a more ambiguous path during the 1930s. The Carnegie Desert Laboratory witnessed slight expansion at first, but by the middle years of the decade the Carnegie Institution began to reconsider the place of the Tumamoc Hill facility in its scientific endeavors. Faced with decreased revenues and no longer committed to ecological research as a fundamental aspect of its research program, the institution soon viewed the Tucson outpost as expendable. They closed the Desert Lab in 1940, transferring the site to the U.S. Forest Service for use by the Southwest Forest and Range Experiment Station. In contrast, the University of Arizona established the Laboratory of Tree-Ring Research on the Tucson campus in 1937. Funding for the laboratory remained meager, but the existence of such a facility at least suggested the possibility of later expansion.[19]

The continued growth of institutions and the scientific community was paralleled during the 1930s by the emergence of the federal government as a major factor in the development of science in the American

Southwest. Although federal agencies had long been visible as part of the region's scientific establishment, the decade of the Great Depression magnified the federal presence significantly. The U.S. Forest Service, for example, chose Tucson as headquarters for the Southwest Forest and Range Experiment Station in 1930. This administrative facility would coordinate the activities of the Santa Rita Range Preserve near Tucson, the Jornado Range Preserve near Las Cruces, and the two units at the Fort Valley Experiment Station near Flagstaff. Equally significant, the Forest Service chose Tucson largely because of the agriculture program at the University of Arizona, which pursued similar research activities. Cooperation between the Forest Service and the university would provide greater efficiency and coordination of various research efforts.[20]

Although the Southwest enjoyed less dramatic benefits from the New Deal than other areas of the nation, scientists in the region found themselves increasingly involved with federal projects. Public Works Administration funds arrived on the University of Arizona campus in 1934 and were used for construction projects, including a new science building. Agricultural scientists played an increasingly important role in the region through their association with the Soil Conservation Service. Clearly focusing on resource management and the application of scientific concepts, the service attempted to improve southwestern agriculture at several levels. John Collier's "Indian New Deal" also brought agricultural scientists from the Civilian Conservation Corps and the Interior Department to the Southwest in a moderately successful effort to improve agriculture by controlling soil erosion and other problems.[21]

Collier's new perspective on Native Americans led to significant changes in medicine on the reservations of the Southwest. In addition to expanding the number of hospitals and physicians serving the native population, his policies brought Native Americans more fully into the health care system. On the Navajo reservation, for example, Collier encouraged Bureau of Indian Affairs physicians and nurses to recognize the value of traditional healing. The Navajo used a variety of medicinal plants to treat symptoms and employed ceremonial activities to deal with the underlying causes of maladies. Bureau health care personnel increasingly cooperated with healers during the 1930s, recognizing that the psychological value of traditional ceremonies was a worthwhile complement to the physiological effect of modern medicine. In turn, many Navajo healers became adept at recognizing serious medical problems and referred their patients to government hospitals for treatment. Efforts were also made to recruit Native Americans as health care professionals. During the summer of 1934, BIA nursing supervisor Elinor Gregg conducted a month-long institute in Santa Fe to train young Navajo women as nurses' aides. Nearly a hundred women attended the institute to learn basic medical care and more specialized treatment of such diseases as trachoma, after which they returned to the reservation to apply their knowledge. Similar programs continued throughout the decade.[22]

Among the most dramatic scientific contributions from the New Deal, however, were those associated with southwestern archaeology. The region had, to be sure, been a central focus of American archaeology for decades. When federal officials began to develop "relief archaeology" as a component of such agencies as the Civil Works Administration, the Federal Emergency Relief Administration, and the Works Progress Administration, the Southwest's archaeological riches attracted attention. Ironically, the region's status as an archaeological bonanza minimized federal spending. Because sites in the Southeast were less well explored and, thus, more labor intensive to excavate, New Deal archaeological projects tended to focus in that region. Southwestern sites were both better known and easier to excavate and therefore failed to offer the relief opportunities that were the focus of New Deal archaeology.[23]

Nonetheless, the Southwest benefited from federal spending for archaeological projects. Funding for such activity began under Civil Works Administration authority during the winter of 1933–34 and quickly led to excavations at Wupatki Pueblo by Harold S. Colton of the Museum of Northern Arizona and at Tuzigoot Pueblo by Byron Cummings and staff of the Arizona State Museum. Federal Emergency Relief Administration funds were also available for this latter project. Other early examples of federally funded archaeological work included excavations at Montezuma Castle in Arizona and stabilization efforts at Aztec Ruin in New Mexico.[24]

As federal relief efforts expanded during the decade, southwestern archaeology continued to enjoy increased funding. FERA support enabled the University of New Mexico and the Museum of New Mexico to conduct five expeditions in the Chaco Canyon area in 1935. Three years later, crews from these institutions and the Laboratory of Anthropology in Santa Fe expanded their work in Chaco Canyon with WPA support and cooperation of laborers from the National Park Service, the Civilian Conservation Corps, and the National Youth Administration. Similar activity characterized the 1930s in Arizona, although that state's archaeological research was marked by an even greater degree of institutional cooperation. Projects funded by federal agencies were generally coordinated by the University of Arizona, but involved the Museum of Northern Arizona, Gila Pueblo, and the Pueblo Grande Laboratory, as well as local and state agencies. In extreme west Texas, Sul Ross College received a WPA grant for work in Presidio County, where a small group of faculty and students excavated village sites and mounds between July 1938 and June 1939. Although federal funding of such projects generally remained modest in the Southwest, the region's archaeologists played an important role in the growing federal presence in the science of the region.[25]

Institutional growth in the Southwest provided the base for a larger and more diverse scientific community in the 1930s. The scientific population listed in *American Men of Science* rose to 169 in 1933 and 215 five years later. Arizona continued to claim the largest number of scientists, primarily because of the dominant status of the University of Arizona, but by the late 1930s New Mexico's share of the region's scientific population had grown dramatically (table 4.1). The scientists who comprised this larger population were even better educated than their predecessors. Well over 90 percent were college graduates and by 1938 nearly 60 percent held doctoral degrees. At the undergraduate level, these scientists had attended many of the leading schools in the United States, but increasingly they had attended schools in the American West. The University of Arizona emerged as a leading producer of southwestern scientists during the 1930s, while California joined such schools as Chicago, Indiana, and Illinois among undergraduate institutions (table 4.4). Fewer changes were evident at the doctoral level. Chicago, Wisconsin, and California remained among the leading producers of science doctorates at both the regional and national level, although such western schools as Stanford, Colorado, and Arizona had begun to appear as occasional suppliers of science doctorates (table 4.5). Disciplinary distributions revealed an interesting change as well. The traditional disciplines among southwestern scientists (agriculture, botany, geology, chemistry) remained strong in the early 1930s, but their dominance had decreased by the end of the decade. Archaeology, astronomy, mathematics, and psychology all grew dramatically (table 4.2), reflecting institutional factors. Archaeologists and astronomers benefited from the growth of research institutions, while mathematicians and psychologists found employment as faculty members of the growing universities in the region. Among starred scientists, astronomers claimed nearly half the *American Men of Science* listings, a clear indication of the importance of the Lowell, Steward, and McDonald Observatories (table 4.3).

Scientists in the American Southwest during the 1930s enjoyed a wide range of employment opportunities. Universities continued to offer many options, but the private sector also offered employment. From his base in Hobbs, New Mexico, geologist Frank B. Conselman coordinated various efforts for the Gulf Oil Corporation in the southeastern part of the state. Conselman, a native of New York, had earned his doctorate at the University of Missouri in 1934 and had served with the Missouri Geological Survey before joining Gulf Oil in 1935. Government employment for scientists became increasingly important during the 1930s. Soil specialist Donald S. Hubbell (Ph.D., Iowa State College, 1932) joined the staff of the Navajo Experiment Station of the USDA in 1934, becoming director of the Gallup facility the following year. The increased interest in scientific forestry also led to positions for scientists. Barnard A. Hendricks became associate range examiner for the Southwest Forest and Range Experiment Station in 1929, expanding his activity after Tucson became the headquarters station for the agency. Forest pathologist Lake S. Gill held a senior administrative position at the USDA Albuquerque office, making excellent use of his Yale Ph.D. in forestry and his several years of experience with the Agriculture Department. The U.S. Bureau of Mines hired such individuals as metallurgist Franklin S. Wartman of Tucson, while the National Park Service employed scientists such as ecologist Donald E. McHenry (Grand Canyon National Park) and archaeologist Dale S. King (Southwest Monuments).

As the decade of the 1930s ended, science in the American Southwest presented a very different image than it had at the beginning of the twentieth century. Scientists working in the region no longer restricted themselves to collecting specimens and artifacts for study elsewhere in the nation. Throughout the first few decades of the century, they increasingly concerned themselves with establishing the institutional base that would allow southwestern scientists to emerge from the colonial status that had long characterized science in the region. Of equal importance, the arrival of new scientists to support the Southwest's scientific institutions led in the 1920s to a sense of community. The region remained sparsely populated, with population centers separated by long distances. As shown by the important role played by the Southwestern Division of the American Association for the Advancement of Science, however, the region's scientists increasingly viewed themselves as members of a regional scientific community. Support for their scholarly activity was rarely generous, but the numerous opportunities for employment brought a wide variety of scientists to the region. As federal support for science grew during the New Deal era, the Southwest's scientific population increased and diversified. Although the region's natural resources remained an important focus for science in the Southwest, by the end of the 1930s the scientific community represented a microcosm of the nation's scientists. An important foundation had thus been established for future growth that would realize the potential of science in the American Southwest.[26]

Chapter 5 Leading Women Scientists, 1920–1950

One of the more productive components of the New Western History has been the reconstruction of women's roles in the settlement and growth of the American West. Historians have shown that women actively participated in the frontier experience and contributed in previously unrecognized ways to the region's growth. Women scientists, however, have rarely been included in the reevaluation of the West's female population. Although the role of women in the growth of American science has received significant attention, such studies have generally focused on the eastern half of the nation with occasional discussions of California.[1] A demographic analysis of leading women scientists employed in the Southwest provides valuable insight into an important group in the intellectual history of the region. The professional characteristics of these women, especially when compared with those of their male colleagues in the Southwest and with their female colleagues throughout the nation, help to reveal the specific role women played in the establishment of the region's science.

Biographical data concerning women scientists in the American Southwest were collected from the *American Men of Science* volumes published in 1921, 1927, 1933, 1938, and 1949.[2] Although this publication remained the standard biographical directory for the nation's scientists, male and female, during the first half of the twentieth century, it tended to under-represent women scientists because of their concentration in the lower academic ranks and research positions. To gain inclusion in these volumes, therefore, women needed particularly noteworthy credentials. Southwestern women listed in *American Men of Science* represented an elite group.

The region's women scientists faced a difficult professional environment, as the American scientific community offered limited opportunities for women before World War II. Although women gained access to major doctoral programs as early as the turn of the century, they rarely found employment at the academic levels their training would suggest. Instead, the leading women scientists continued to teach and conduct research at women's colleges, while other women found employment in the increasingly prevalent coeducational and land-grant colleges, in a few government agencies, and in a small number of industrial firms. Outside of the elite women's colleges, however, women scientists tended to be concentrated in the lower ranks or positions with little chance for significant advancement. They frequently served as voluntary or poorly paid assistants in laboratories despite holding graduate degrees from major universities.[3]

Women scientists contributed to their own "invisibility" by accepting overqualification as the price for recognition from and admission to the wider scientific community. Historian Margaret Rossiter described this internalized double standard as the "Madame Curie strategy," which took on greater visibility after Marie Curie's fund-raising visit to the United States in 1921. Although newspaper articles discussed her life in great detail and applauded her for "opening doors" for women scientists, the impact on her American colleagues proved detrimental. Very few American scientists (male or female) were the intellectual equals of Madame Curie, yet in order to be considered for an appropriate academic post an American woman scientist had to convince her male colleagues that she had reached Curie's level of achievement. Despite their higher educational qualifications (those listed in *American Men of Science* held the Ph.D. in significantly higher proportions than did their male colleagues), women scientists generally remained mired in the lower ranks of their academic disciplines.[4]

The prospects for women scientists in the United States were not totally bleak. Throughout the first few decades of the twentieth century, the concept of "women's work" in science led to opportunities in specific fields. Chemist Ellen Swallow Richards, for example, applied her training at Vassar and the Massachusetts Institute of Technology to the creation of home economics as a scientific field by 1910. Demonstrating the importance of science to many facets of homemaking, Richards's work led to a discipline that became particularly valuable to the emerging land-grant schools and extension services in the Midwest and West. Other disciplines that provided women scientists with professional opportunities included nutrition (as in home economics, women chemists tended to dominate this field), hygiene, and to a lesser extent botany, medicine, and anthropology. In these disciplines women could advance even if salaries remained consistently below those of men with similar rank. Coeducational colleges and universities also provided women scientists with administrative opportunities in the office of dean of women. Campus officials frequently added such positions to the duties of female faculty members regardless of their background or qualifications.[5]

As women scientists attempted to gain access to legitimate roles in the American scientific community, science in the Southwest grew steadily into a position of respectability. Despite this growth, women scientists became visible in the region very slowly.[6] The first two women listed in *American Men of Science* appeared in the 1921 volume, displaying many of the characteristics of their female colleagues throughout the nation. Edna Mosher, a native of Canada, earned an undergraduate degree at Cornell University in 1908, after which she studied entomology at the University of Illinois. A doctoral graduate of 1915, she taught at Ohio State University and the University of Illinois before accepting a position as professor of biology at the University of New Mexico in 1919. Her appointment to a senior position represented an unusual degree of recognition for a woman scientist, but her selection as dean of women in 1920 more closely followed the academic practices of the period. She held these two positions until 1923, when she accepted a professorial post at Adelphi College in Garden City, New York.

If Mosher's career in New Mexico suggests that opportunities for women scientists were greater in the region than in the nation as a whole, the experiences of Edith Shreve provide an alternative image. A graduate of the University of Chicago with degrees in physics and chemistry, she was a physics instructor at Goucher College when she met botanist Forrest Shreve. After their marriage in 1909, Shreve joined her husband in his work at the Carnegie Desert Laboratory near Tucson, quickly establishing herself as a plant physiologist. By 1914, for example, she had published a classic study on palo verde transpiration which remains part of the standard literature. She accompanied her husband on field expeditions but generally worked independently on her own research topics.[7] Despite her recognized ability Edith Shreve served as a "voluntary investigator" for the Desert Laboratory, a situation imposed on many women scientists by strictly enforced nepotism rules.

Over the next dozen years the number of listed women scientists in the Southwest remained stable at four. While Shreve continued her research in plant physiology,[8] other women arrived in the region to pursue various studies. Botanist Frances Louise Long briefly worked at the Desert Lab in the late 1920s (before moving to the Carnegie Institution's Santa Barbara facility), while biologist Helen Murphy (University of New Mexico) and chemist Lila Sands (University of Arizona) accepted faculty appointments in 1923 and 1924, respectively. The 1933 cohort included two new names, both of whom held positions at the University of Arizona. Bacteriologist Mary Estill Caldwell, a recent recipient of a University of Chicago doctorate, was assistant professor of biology. Born in Ohio, she had earned undergraduate and master's degrees at the University of Arizona, after which she taught biology for the institution as an instructor from 1919 to 1924. She married zoologist (and Arizona faculty member) George T. Caldwell in 1925. A Rockefeller Foundation Fellowship for 1929–30 enabled her to pursue doctoral study. Margaret Cammack Smith came to Tucson in 1925 as an associate professor of home economics after teaching at Columbia Teachers College and earning a doctorate from Columbia University. A specialist in nutrition, her research interests paralleled those of her husband, Arizona agricultural chemist Howard Smith, whom she married in 1927. By 1928 she had been promoted to the rank of professor of nutrition and nutrition chemist at the university.

The women scientists employed in the Southwest during the 1920s and early 1930s, although small in number, nonetheless provide intriguing insights into the role such women played in the establishment of a scientific community in the region. As was the case with their male colleagues, most of the women were born in the Midwest and attended such leading

Mary Estill Caldwell served as head of the University of Arizona bacteriology department from 1935 until 1956. (University of Arizona Library Special Collections; University of Arizona Biographical File, neg. no. N-1756)

undergraduate and graduate institutions as Cornell University, the University of Chicago, and the University of Illinois. These schools were the primary sources of women scientists in the nation, indicating that southwestern women scientists shared many of the professional characteristics of the national population.[9]

The Southwest's women scientists displayed other characteristics of their sisters elsewhere. The concept of "women's work" remained quite evident in the employment histories of these women. Such "feminized" fields as nutrition and home economics were well represented, as were the slightly less feminized specialties of botany and biology. The new discipline of bacteriology emerged as a field of interest to women, at least in part because of its connection with various health reforms of the early twentieth century.[10] The "Madame Curie strategy" was also evident, as indicated by the high proportion of Ph.D. recipients (Edith Shreve was the only woman scientist without this degree).

During this early period in the Southwest's scientific growth women appear to have gained promotion more easily than colleagues elsewhere. Women enjoyed greater opportunities for advancement in western colleges and universities throughout the twentieth century. A study published in *Science* in 1912 indicated that women were more than twice as likely to hold positions in state colleges and universities west of the Mississippi River than in those east of the river. Although these women tended to be employed in fields other than science and at the rank of instructor, more than 10 percent (primarily in western schools) held the rank of professor.[11] Among southwestern women scientists, Mosher, Murphy, and Smith progressed rapidly to senior positions, while Caldwell and Sands earned posts as assistant professors more quickly than individuals with similar credentials in other regions. Admittedly, the small population makes generalizations suspect, but opportunities for women scientists clearly existed in the Southwest of the 1920s and early 1930s.

The 1938 edition of *American Men of Science* included ten women scientists from the Southwest. As had been the case earlier, these women displayed a strong midwestern connection. Sixty percent of the 1938 cohort were born in the Midwest (table 5.1) and half of them earned undergraduate degrees in the region (table 5.2). Unlike their female colleagues elsewhere, southwestern women scientists were overwhelmingly educated in coeducational institutions; Goucher College, with one alumna, was the only women's college represented among the 1938 population. In addition to the midwestern concentration, a secondary undergraduate axis of Arizona and California appeared among these women scientists. Stanford University and the University of California each produced one graduate, while the University of Arizona claimed two. Eighty percent of these scientists held the doctoral degree, a significantly higher figure than that of the region's scientific population as a whole. Major programs at the University of Chicago, Columbia University, and the University of California were represented, as were slightly less well established programs at such schools as Ohio State University and the University of Nebraska (table 5.3). Graduates appear to have continued their education at the same school in most cases. Of those who pursued a master's degree (nine

Table 5.1 Birthplaces of Women Scientists, 1920–1950

State or Country	*American Men of Science* Volume				
	1921	1927	1933	1938	1949
New York				1	1
New Jersey					2
Maryland				1	
North Carolina					1
Tennessee					1
Arkansas					1
Oklahoma					1
Ohio			1	2	1
Michigan	1	1	1	1	1
Illinois				1	1
Missouri					1
Iowa					1
Nebraska		2	1	1	1
Minnesota			1	1	3
Colorado					2
California				1	
Canada	1				
Mexico				1	1
Scotland		1			

of the ten), all but three remained at their undergraduate institution. Of the eight who earned the doctorate, half did so at their undergraduate school. The two University of Arizona graduates remained in Tucson for their master's degrees and then moved to the University of Chicago for doctoral work. In terms of their educational background, the region's women scientists possessed credentials little different from their male colleagues.

The "women's work" perspective continued to characterize the region's

Table 5.2 Undergraduate Institutions of Women Scientists, 1920–1950

Institution	American Men of Science Volume				
	1921	1927	1933	1938	1949
University of Chicago	1	1	1	1	3
University of Arizona			1	2	3
University of Nebraska		2	1	1	1
University of Minnesota			1	1	1
University of California				1	1
Cornell University	1	1			
University of New Mexico					2
Stanford University				1	1
Drake					1
Goucher				1	
Illinois State Normal					1
Miami (Ohio)				1	
Mount Holyoke					1
New Jersey College for Women					1
North Texas State College					1
Northwestern University				1	
Pittsburgh					1
University of Washington					1

female scientists as the decade of the 1930s ended (table 5.4). Such fields as medicine, nutrition, anthropology, and botany dominated the specialties these scientists claimed. Although two chemists appear among the 1938 population, one of them (Louise Otis of the Arizona Agricultural Experiment Station) focused on nutrition research. Women scientists found employment at several different institutions (only the University of Arizona employed more than one), ranging from colleges and universities to independent research facilities such as the Gila Pueblo Archaeological Foundation and the Carnegie Desert Laboratory (table 5.5). Data concerning academic employment suggest that women continued to fare better

Table 5.3 Doctoral Institutions of Women Scientists, 1920–1950

Institution	*American Men of Science* Volume				
	1921	1927	1933	1938	1949
University of Chicago			1	2	5
Columbia University			1	1	2
University of Nebraska	1		1	1	1
Cornell University	1				1
Stanford University				1	1
University of Arizona					1
University of California				1	
Clark University					1
Johns Hopkins University					1
University of Illinois	1				
University of Iowa					1
University of Minnesota		1			
Northwestern University				1	
Ohio State University				1	
University of Texas					1

in the Southwest than their female colleagues elsewhere. Of the five who held academic positions, two were full professors in 1938 while the other three were divided equally among the instructor, assistant professor, and associate professor ranks. Women who held research positions tended to be concentrated at the junior staff level, a situation common for women scientists in the late 1930s.

Southwestern women scientists contributed to their professions in a number of different ways. Edith Shreve, for example, continued to volunteer her services to the Desert Laboratory, pursuing her own research interests while assisting her husband in his. Another husband-wife team accomplished their work at the Arizona Agricultural Experiment Station in Tucson. Nutritionist Margaret Smith and her agricultural chemist husband Howard investigated the mottling of human tooth enamel during the

Table 5.4 Specialties of Women Scientists, 1920–1950

| | *American Men of Science* Volume | | | | |
Discipline	1921	1927	1933	1938	1949
Mathematics				1	
Physics					2
Chemistry		1	1	2	1
Biology	1	1		1	1
Bacteriology			1	1	1
Botany		1			2
Plant Physiology	1	1	1	1	1
Medicine				1	2
Psychology					2
Nutrition			1	1	2
Anthropology				2	2
Archaeology					1
Geography					1
Home Economics					1

1930s and determined that excessive fluorine in water supplies caused this condition. They developed an effective process to remove this element in 1945, after which they became increasingly involved in the debate concerning the fluoridation of water supplies in the United States. Their work was often cited by opponents of such efforts, although by the mid-1950s the Smiths had accepted the value of fluoridation under appropriate safeguards.[12]

Another woman scientist found the Southwest a land of exceptional opportunity. Mary Caldwell's career at the University of Arizona progressed rapidly during the 1930s, leading to her appointment as head of the bacteriology department in 1935 and her promotion to full professor two years later. She served as department head for more than twenty years, after which she continued as professor of pharmacology and research pharmacologist in the College of Pharmacy. In a period of few such opportuni-

Table 5.5 Employers of Women Scientists, 1920–1950

	American Men of Science Volume				
Employer	1921	1927	1933	1938	1949
University of Arizona		1	3	3	7
Carnegie Desert Laboratory	1	2	1	1	
University of New Mexico	1	1		1	1
Los Alamos Scientific Laboratory					3
New Mexico Highlands (Las Vegas)					2
Arizona Agricultural Experiment Station				1	
Arizona State Teachers College (Flagstaff)					1
Arizona State Teachers College (Tempe)					1
Gila Pueblo				1	
U.S. Department of the Interior				1	
Museum of New Mexico					1
National Research Council					1
New Mexico Military Institute				1	
Private sector					1
Texas Western College					1
Retired				1	

ties for women scientists, Caldwell's success was sufficiently unusual to attract the attention of later scholars. Caldwell was the only southwestern scientist discussed in Rossiter's *Women Scientists in America*. She suggested that Caldwell's promotion was eased by her two Arizona degrees, which provided senior colleagues with additional knowledge of her abilities.[13] Although not mentioned by Rossiter, Caldwell's research interests in tuberculosis undoubtedly contributed to her colleagues' high esteem.

The greatest opportunities for the region's women scientists appeared in anthropology. Combining as it did the "women's work" concept and the traditional importance of the Southwest to the discipline, anthropology provided opportunities during the 1930s even for individuals trained in other fields. Physician Sophie B. Aberle (Ph.D., Stanford, 1927; M.D.,

Yale, 1930) arrived in Albuquerque in 1935 as the superintendent of Pueblo
Indians for the Bureau of Indian Affairs. In addition to her administrative duties, she conducted research concerning the vital statistics of the
Pueblo Indians and the growth and development of their children. The
federal government's growing concern with the medical problems facing
Native Americans during the 1930s provided Aberle with significant support for her endeavors.[14]

Of greater moment for anthropology itself, two figures stand out among
the women scientists of the late 1930s. Isabel Kelly (Ph.D., University of
California, 1932), a student of noted anthropologist Alfred Kroeber and
sociologist Mary Roberts Coolidge, served two years as a National Research Fellow at the Laboratory of Anthropology in Santa Fe. In 1937
she joined the Gila Pueblo Archaeological Foundation in Globe, Arizona,
for which she excavated the important Hodges Site north of Tucson. Her
endeavors provided specific information about the Hohokam culture and
paralleled other investigations in the state. Although Kelly was most famous for her later work in western Mexico during the 1940s and 1950s,
her contributions to southwestern archaeology were no less significant.[15]

Even more noted for her anthropological research in the region, Florence Hawley also represents one of the few native southwesterners to
make an early contribution to science. Born in La Cananea, Mexico, in
1906, Hawley pursued her undergraduate and master's degrees at the University of Arizona where she studied with respected archaeologist Byron
Cummings and dendrochronologist A. E. Douglass. One of the first women
to earn a doctorate in archaeology (University of Chicago, 1934), she applied the new technique of tree-ring dating to determine the age of the
Chetro Ketl ruin in Chaco Canyon and later extended dendrochronological
research to the Mississippi Valley. She accepted an assistant professorship at the University of New Mexico in 1935 and held a joint appointment at the University of Chicago from 1937 until 1940.[16] Already a leading figure in archaeology by the mid-1930s, Hawley's presence aided in
the establishment of the University of New Mexico as a second focus for
southwestern archaeology, joining the well-established program at the
University of Arizona.

World War II brought about dramatic growth of the Southwest's scientific community and an increase in the number and status of women scien-

tists in the region. The number of southwestern women scientists listed in *American Men of Science* reached nineteen by 1949, displaying noticeably different demographic characteristics compared with their predecessors of the late 1930s. Although the Midwest remained the region of birth for the largest number of these scientists (nearly 50 percent), the Middle Atlantic states and the South also claimed large numbers (table 5.1). The Midwest and Middle Atlantic states had long been the origin of the majority of the region's scientists, but the South appears as a leading source only among women scientists. The educational backgrounds of the region's women scientists also changed during the decade. At the undergraduate level, schools in the Southwest and on the Pacific coast supplanted the midwestern colleges and universities as the leading suppliers of these scientists (table 5.2). The University of Chicago claimed three alumnae, but the only other schools with multiple graduates were the University of Arizona with three and the University of New Mexico with two. The presence of these latter two institutions among leading producers of science undergraduates provides indirect evidence of the growth of science in the region during the preceding decades.

Women scientists in the Southwest continued to represent a better-educated population than their male colleagues. All but one of the nineteen held postgraduate degrees, with more than 84 percent recipients of the Ph.D. or M.D. degree. The corresponding figure for the region's scientific population as a whole was slightly less than 60 percent. The University of Chicago remained the primary source of science doctorates among women scientists, producing a third of all degrees, but other leading science programs included Columbia, Johns Hopkins, and Cornell universities (table 5.3).[17] Although the region's universities were slowly adding doctoral programs, only the University of Arizona produced a Ph.D. among this population, awarding a botany degree to Kittie F. Parker in 1946. Parker had been in Tucson since 1937 when her husband, Kenneth, accepted a position with the Southwest Forest and Range Experiment Station of the U.S. Forest Service. She served as laboratory assistant and instructor until 1949, when she became an assistant professor of botany and curator of the university's herbarium. Unlike their colleagues of a decade before, the 1949 cohort often chose not to remain at their baccalaureate institutions for graduate study. Ten of the seventeen who earned master's degrees remained at their undergraduate schools, but only four of the fifteen recipients of the Ph.D. earned their doctorates from the schools granting their bachelor's degrees.

Employment patterns show less remarkable change during the decade, although certain developments are noteworthy (table 5.5). Academic appointments became even more dominant in the late 1940s, with nearly 70 percent of all women scientists employed by the universities and colleges of the region. The University of Arizona remained the leading location of women scientists, employing seven of the nineteen, while New Mexico Highlands University included two women scientists among its faculty. The ranks these women held indicate a continued distribution which was more advanced than that which characterized women scientists nationwide. Nearly half of the women academic scientists held full professorships and only two remained below the associate professor level.

Two phenomena emerge to explain this concentration in higher ranks. In many cases, promotions occurred in a relatively rapid fashion. Nutritionist Ethel M. Thompson, for example, joined the University of Arizona faculty as an assistant professor in 1938. After receiving her Ph.D. from Columbia University in 1940, she advanced to associate professor, followed by a professorship in 1942. In disciplines less identified as "women's work," southwestern women scientists had a more difficult path to senior status, although even here they appear to have been more successful than their colleagues elsewhere. Chemist Lila Sands gained promotion to full professor in 1938, after eight years as associate professor, while clinical psychologist Anna Y. Martin assumed the chairmanship of the psychology department at New Mexico Highlands in 1949 after three years as associate professor and dean of women. Southwestern schools also tended to hire women scientists at advanced ranks. Botanist Mary H. Wilde (Ph.D., Cornell, 1942) taught six years at Mount Holyoke College as instructor and assistant professor before moving to Texas Western College in El Paso as associate professor in 1949. New Mexico Highlands University was sufficiently motivated to hire biologist Lora M. Shields that they offered her an associate professorship upon completion of her Ph.D. at the University of Iowa in 1947.

Specializations of the region's women scientists underwent little change between the late 1930s and late 1940s (table 5.4). Concentration in traditional fields appears even more pronounced, with all but three of the nineteen women in fields such as medicine, nutrition, anthropology, psychology, or botany. Two of the three exceptions to this concentration were physicists Elizabeth Graves and Jane Hall, who had joined the staff at Los Alamos Scientific Laboratory in 1943 and 1945, respectively. Graves's husband, Alvin (whom she had married while they were students at the University of Chicago), had been a Los Alamos staff member since 1943 as well, and served as a division leader in the late 1940s.

Although most of the region's women scientists remained at least moderately active in research, two disciplines (medicine and anthropology) attracted the greatest interest. The Southwest's reputation as a haven for those afflicted with respiratory illnesses led to much research on tuberculosis.[18] Mary Caldwell continued her bacteriological studies of the disease, while Edith Shreve joined the Desert Sanatorium and Research Institute in 1940 as a research biochemist. Shreve's new studies paralleled her husband's decreasing activity at the Carnegie Desert Laboratory after the late 1930s. Even before his retirement in 1945, Forrest Shreve had spent less and less time at the laboratory as his health slowly deteriorated.[19] Physician Jean C. Sabine, who had joined the medical research staff at Los Alamos in 1949, investigated the activity of the enzyme cholinesterase in blood disorders such as pernicious anemia.

The field of anthropology, however, continued to represent the discipline with the greatest opportunities for women scientists in the Southwest. By the late 1940s, Florence Hawley gained promotion to associate professor at the University of New Mexico, where she continued her research concerning the region's native inhabitants. This work attracted national interest and led to her appointment to the elections committee of the American Anthropological Association in 1948.[20] Hawley also devoted significant time to her teaching responsibilities, combining classroom activities and fieldwork at all levels. One of her first students in Albuquerque was Bertha P. Dutton, who earned undergraduate and master's degrees under Hawley in the mid-1930s. Dutton secured a staff position with the Museum of New Mexico in 1936 and soon became curator of ethnology, the first of many administrative positions she held at the museum.[21] The University of Arizona employed the third woman anthropologist in the region. Clara Lee Tanner had been one of the first archaeology graduate students at Arizona in the late 1920s (the other two were Florence Hawley and Emil W. Haury) and became an instructor at the

Tucson school after earning her M.A. degree in 1928. A specialist in southwestern Indian art, Tanner became an assistant professor in 1935.[22]

In addition to their success in terms of promotion and status at their home institutions, women scientists in the Southwest also gained recognition on the national and state levels. Both Hawley and Dutton, for example, were fellows of the American Anthropological Association, while geographer Agnes Morgan Allen (Arizona State Teachers College in Flagstaff) held a position on the executive board of the National Council of Geography Teachers. Sophie Aberle retired as superintendent of Pueblo Indians in 1944 to accept an administrative position in the Division of Medical Sciences of the National Research Council, remaining in Albuquerque while discharging her responsibilities. Two University of Arizona faculty members held important state positions in their disciplines. Nutritionist Ethel Thompson served as chairman of the Arizona Nutrition Council from 1944 to 1945 and was elected president of the Arizona Dietetic Association the following year. Mary Caldwell remained a member of the State Board of Examiners in Basic Sciences and became president of the Arizona Public Health Association in 1946.

Women scientists in the American Southwest from the 1920s through the 1940s represent an intriguing subpopulation of the region's scientific community. Never a large group, they nonetheless displayed many of the characteristics that defined women scientists in the United States during the first half of the twentieth century. As did their female colleagues elsewhere in the nation, they tended to specialize in those fields that were defined as "women's work," such as nutrition, home economics, and to a lesser extent psychology, anthropology, botany, and medicine. Another similarity was the generally higher educational levels among women scientists in the Southwest, as compared to their male colleagues. The research contributions of such figures as Hawley, Smith, and Shreve also indicate a desire to perform at high scholarly levels. The "Madame Curie strategy" clearly applied to the region's women scientists as it did to those elsewhere in the United States.

There exist, however, important differences that also characterize the Southwest's women scientists. Especially in higher education, these scientists appear to have been more successful in gaining senior status than their colleagues elsewhere. Southwestern colleges and universities, because they were land-grant and coeducational, may well have represented a particularly fertile growth medium for this population. These schools were not only willing to hire faculty members in fields such as nutrition and home economics, but also provided similar opportunities in agricultural experiment stations. The growth of these institutions in the 1920s and 1930s provided more opportunities for faculty appointments (for both women and men), leading the region's administrators to hire female faculty more readily than their administrative colleagues in the East and Midwest.[23] For a variety of reasons, women scientists in the Southwest enjoyed opportunities unavailable to their female colleagues in other areas of the nation.

At mid-century, women scientists in the Southwest had established noteworthy individual reputations and had played a visible role in the growth of science in the region. Although a small percentage of the scientific community, these women nonetheless held important positions at their home institutions and earned state and national recognition among their peers. They participated in the institutional growth of the region's science and contributed to the advance of their disciplines. A hitherto invisible aspect of the region's development, these women scientists represent an important addition to the history of both women and science in the American Southwest.

Chapter 6 The Impact of World War II

Historians have long identified World War II as the key dividing point of twentieth-century America. Political, economic, diplomatic, and social histories are largely defined in terms of pre- and postwar developments marked by dramatic discontinuities. From the perspective of most historians, the war created the Southwest's science, as symbolized by the establishment and growth of Los Alamos. From this base, the postwar expansion of science in the region paralleled economic and political developments and created the modern Southwest. No longer did the region's scientists appear isolated from the national scientific community. Leaders of American science could be found in New Mexico as well as in Massachusetts.

The impact of the war on science in the American Southwest certainly appears dramatic. As indicated by listings in the *American Men of Science* volumes published in 1938 and 1949, the region's scientific community grew explosively. From 215 individuals listed in the last prewar volume, the region's scientific population grew to 504 in the first postwar volume. Equally dramatic and seemingly even more indicative of war-induced changes, a new disciplinary concentration was evident among the region's scientists in the late 1940s. Before the war southwestern scien-

tists tended to pursue work in such fields as geology, biology, and archaeology, thus taking advantage of the region's natural resources. After the
war the new glamour field among southwestern scientists was physics,
as was the case throughout the United States. The disciplinary concentration was paralleled by a geographical concentration as well, with nearly
a quarter of all southwestern scientists listed in the 1949 *American Men
of Science* volume identified with the atomic research programs at Los
Alamos and in Albuquerque.

A closer examination of the region's science and scientists during this
period reveals a more complex and intriguing situation than is usually
portrayed. Although war-related changes in the region's science were profound, they were responsible for the *development*, rather than the *creation*
of southwestern science. The physicists at Los Alamos represented one
chapter in the story of the region's science, but other chapters also
emerged, several of which were continuations of those from earlier years.
By the end of the decade of the 1940s, the Southwest was thus characterized by two scientific traditions. One was based on the new science of
large budgets and research teams, symbolized by Los Alamos, while the
other continued to pursue new discoveries based on the region's status as
a natural laboratory. These two traditions complemented each other and
defined the new scientific world of the American Southwest.

Among the more significant displays of continuity during the period was
the continued role of science in the region's economic growth. An especially noteworthy example of this activity was the contribution made by
scientists in the development of the Arizona citrus industry. Farmers in
the Salt River Valley and in the Yuma area had begun growing citrus (primarily oranges) as early as the 1890s, but they continued to operate on a
small scale for many years. The key change in the state's citrus industry
was the expansion of grapefruit cultivation during the late 1920s to take
advantage of the relatively untapped market for this product. When the
new plantings began to mature in the mid-1930s, Arizona production figures rose dramatically, exceeding 2.7 million boxes in 1937 and approaching 4.1 million boxes in 1943.[1]

Although the development of new grapefruit varieties was largely the
work of citrus research stations in California and Florida, scientists in
Arizona also applied their knowledge to this important addition to the

state's economy. The Yuma branch of the Arizona Agricultural Experiment Station purchased a 160-acre plot in 1919 as a citrus farm, primarily to study the impact of growing conditions on production. The U.S. Department of Agriculture's Sacaton field station began the study of grapefruit in the same year, expanding their investigations during the next decade. Most of the scientists involved in citrus research during the 1930s focused on the commercial prospects of Arizona grapefruit. Horticulturists from the experiment station examined methods of storage in the Phoenix area (the commercial center of the state's grapefruit industry) not only to learn the optimum temperature and humidity for storage but also to determine when grapefruit should be harvested. Experiment station staff investigated growth processes in grapefruit to discover the impact of irrigation and fertilization practices on maturity and quality of the crop. Conducting research on the Yuma farm and in the Salt River Valley groves, these scientists made direct contributions to the expanding citrus industry in Arizona.[2]

As the grapefruit industry grew during the 1930s, other Arizona scientists examined the food value of the new product. Nutritionist Gladys Hartley Roehm of the University of Arizona, for example, recognized the importance of grapefruit to the state's economy in the early 1930s and began to analyze the fruit. She focused on a study of vitamins B and G, complementing significant research already accomplished concerning vitamin C content. Analyzing Marsh grapefruit grown on the Yuma farm, Roehm conducted a series of experiments on laboratory rats. Based on the weight gain of these rats during an eight-week experimental regimen, grapefruit pulp was deemed to be an "excellent" source of vitamin G (as was the peel) but a poor source of vitamin B. During the early 1940s a team of University of Arizona nutritionists worked with members of the experiment station staff to determine the impact of nitrogen content on the quality of grapefruit. They analyzed Yuma and Phoenix area produce, determining that lower nitrogen levels resulted in higher quality crops. Later studies explored the impact of various preparation techniques on several different citrus varieties.[3] Whether examining ways to increase production or analyzing the nutritional value of the crop, scientists played a central role in the growth of the Arizona citrus industry, continuing a long tradition of applying science to agriculture in the Southwest.

In other respects, however, discontinuities represent a valuable focus

of study. Few events in the twentieth century did as much to change
the American West as did World War II. Historian Gerald Nash identified
the war as changing the region from a colony of the East to "the pace-
setting region of the nation" and emphasized that the war both expanded
and diversified the West's economy.[4] The Southwest shared in the various
changes brought about by the war, as indicated by the dramatic population growth between 1940 and 1950. Arizona grew by 50 percent during
this decade; New Mexico grew by nearly 30 percent. The most obvious
war-related change was the military expansion in the region, best shown
by the many air bases established or expanded during the war. Defense
plants were equally visible and contributed significantly to the population and economic growth of the region. Consolidated Vultee established a
large plant in Tucson, while the Phoenix area witnessed the construction
of plants by Goodyear Aircraft Corporation, Alcoa, and AiResearch.[5]

The beginnings of the Cold War also had a major impact on the Southwest's new focus on defense industries. Military installations remained
important, but electronics and other military research and development
activities were more notable. In early 1948, for example, Chicago-based
Motorola decided to locate a new research and development center in
Phoenix. Soon established as the Military Electronics Division, the Motorola plant took advantage of the dry climate and local business support.
Albuquerque enjoyed similar postwar expansion in areas related to atomic
science and related military concerns, while Tucson eventually attracted
a Hughes Aircraft Corporation plant. The attraction of the Southwest for
such facilities established the region as a manufacturing center and laid
the foundation for the explosive growth of the 1950s that established the
Sun Belt phenomenon.[6]

World War II played a major role in creating a new perspective on
American science that focused on government funding. Federal support
for science in the West concentrated in California because of the state's
well-established institutional base, but the Southwest soon began to share
in the financial windfall. The Office of Scientific Research and Development (OSRD) negotiated contracts with such institutions as the University of New Mexico. Among the projects conducted by the Albuquerque
school were those to establish rocket proving grounds in the region and
to test captured German rockets. Even more dramatic, physics department chair E. J. Workman directed a research project at the university

that played an instrumental role in developing the proximity fuse for artillery shells.[7] Although the universities of the Southwest did not have the reputation of such western institutions as Caltech, Stanford, or Berkeley, and thus did not receive the level of funding enjoyed by these California schools, OSRD and other contracts nonetheless established the region as a participant in the new science of World War II.

The most dramatic example of wartime science in the American Southwest was the isolated laboratory northwest of Santa Fe known as Los Alamos. Alerted to the weaponry potential of uranium fission in 1939, military and civilian officials became increasingly intrigued with this topic. By the summer of 1942 the federal government had established a separate research and development program to coordinate the creation of an atomic weapon. Army engineer Leslie Groves, who had recently completed the construction of the Pentagon, assumed the direction of the Manhattan Project in September. Although most of the project's estimated two-billion-dollar cost was absorbed by the production facilities in Oak Ridge, Tennessee, and Hanford, Washington, the outpost in New Mexico would be responsible for the design and construction of the weapons.[8]

Working closely with noted physicist J. Robert Oppenheimer and other leading scientists, Groves selected the Los Alamos site largely because of its isolation. Not only would this remote location simplify security, but it would also concentrate the scientists into their own critical mass. Work at Los Alamos began under the direction of Oppenheimer in the spring of 1943, through a contract signed with the University of California at Berkeley, a noted center for nuclear research. Oppenheimer initially viewed his New Mexico outpost as a small campus-like setting populated by several dozen scientists pursuing theoretical as well as practical questions. This vision evaporated quickly. Los Alamos soon became a major installation, as shown by the large number of Corps of Engineers personnel involved in construction activity throughout the war. The technical and scientific staff, composed of civilians and military personnel, had reached more than a thousand by mid-1943, a number that more than tripled during the next two years. Other personnel swelled the Los Alamos population to more than six thousand by the end of the war. Despite its secret status, the laboratory represented one of the largest scientific facilities in the United States.[9]

Under Oppenheimer's direction, Los Alamos attracted an overwhelming concentration of scientific talent. By the summer of 1945, the Los Alamos scientists had solved numerous theoretical and technical difficulties and had designed uranium and plutonium bombs to use the limited fissionable material available. The Trinity Site test in southern New Mexico on 16 July 1945 proved the feasibility of the implosion technique for the plutonium device. The uranium bomb dropped on Hiroshima (6 August) and the plutonium weapon employed against Nagasaki (9 August) marked the beginning of the end of wartime Los Alamos and led to the announcement of the secret city that had emerged in northern New Mexico.[10]

The end of the war revealed Los Alamos as a central addition to the scientific establishment of the American Southwest, but the future of the facility remained uncertain. Having achieved their goal of an atomic weapon, the hill's population began to return to their prewar lives. By early February 1946, all division leaders had decided to leave the laboratory, following Oppenheimer's departure as director the previous fall. Postwar debates in Washington concerning the place of atomic science and weapons added to the uncertainty. Efforts to maintain the laboratory fell to Norris Bradbury, a leading figure in the Explosives Division during the war who assumed the directorship following Oppenheimer's departure. Although anxious to learn its ultimate fate, the facility remained active. Engineering work on bomb designs continued during the months after the war, while initial research on fusion weapons began. The laboratory was heavily involved in Operation Crossroads, the first postwar atomic weapon test. Beginning in December 1945, Los Alamos scientists and technicians designed and constructed test equipment to measure the impact of an atomic explosion against naval vessels. The tests in June and July 1946 at Bikini Atoll involved the laboratory in a major project and increased the monthly payroll to more than $500,000 at its peak.[11]

Operation Crossroads suggested that Los Alamos had a role to play in the national defense program, but the installation's scientific focus defined it as more than a military base. Bradbury reorganized the laboratory and began to develop Los Alamos as a community rather than an isolated outpost. In July 1946 the laboratory hosted a University Affiliations Conference to discuss the possibilities of research cooperation with universities in the West. Equipment at Los Alamos could be used by graduate students in chemistry, physics, and metallurgy, expanding the work of the New Mexico laboratory beyond weapons research. The key change

The shop area, motor pool, and fire station of Los Alamos Scientific Laboratory in the late 1940s. (Courtesy of the Los Alamos Historical Museum Archives; no. P1989-13-1-1188)

for Los Alamos was the establishment of the Atomic Energy Commission in 1946 and its decision to direct the laboratory beginning 1 January 1947. Federal funding removed much of the uncertainty facing Bradbury and his colleagues, allowing them to begin planning for the expansion of both the laboratory and the community. Recruitment of scientists and engineers became much less difficult and the former secret base became a well-known research facility. As indicated by the 119 individuals from Los Alamos listed in the 1949 *American Men of Science*, the laboratory emerged over the next few years into the leading scientific facility in the American Southwest.[12]

Although uncertainty characterized Los Alamos following the war, few doubted that weapons research would continue. Among the perceived needs was a weapons development site, to be established elsewhere because of the lack of space on the plateau. Toward this end, Z Division was formed during the summer of 1945 from the Ordnance Engineering Division of Los Alamos. Relocated during the spring of 1946 to an abandoned airfield near Albuquerque, Z Division inherited the responsibility for weapons-related research, development, and testing. The Atomic Energy Commission examined the Albuquerque base on several occasions during 1947, ultimately reorganizing the division to increase its readiness and to make it more independent of the theoretical research focus of Los Alamos. Officially defined as a separate branch of the Los Alamos Scientific Laboratory, Z Division (known as Sandia Base) increasingly employed private contractors throughout the United States for the production and assembly of weapons. Although the Sandia reorganization proved effective, the division's intimate involvement with the actual production of atomic weapons troubled the University of California, who held the contract for the administration of Los Alamos. Berkeley officials remained committed to the administration of the New Mexico facility, but asked the Atomic Energy Commission in December 1948 to remove Sandia Base from the contract. By the following fall, Z Division had been transformed into the Sandia Corporation to operate the installation under the management of Bell Laboratories. Nearly 1,800 employees worked at Sandia, a number that more than tripled over the next five years.[13]

Although the Manhattan Project emerged as the key to the Southwest's wartime and postwar science, the region also played a crucial role in another important technical development. Reports of the successful German long-range missile program attracted significant interest among American scientists, who soon developed plans for the capture and study of V-2 rockets. Among these scientists were Gerard P. Kuiper of Yerkes Observatory, Donald Menzel and Fred Lawrence Whipple of Harvard, Fritz Zwicky of Caltech, and James Van Allen of the Applied Physics Laboratory at Johns Hopkins University. Equally important, the U.S. Army Ordnance Department sent teams of engineers across the nation to find the best site for a long-range proving ground. By the spring of 1945, the White Sands region of southern New Mexico was chosen and arrangements were made to obtain German rockets and ship them to the new site. The Corps of Engineers constructed launch pads, blockhouses, and other buildings during the summer, enabling the first launches of small developmental rockets during the fall.[14]

The key to the beginnings of the American rocket program was the arrival of Werner von Braun and his colleagues from Germany in September of 1945. Von Braun and the initial group of German scientists arrived in El Paso the following month and began organizing for their research at

White Sands. Components for a hundred V-2 rockets were already stored in the desert waiting for scientists and engineers to assemble and launch the devices. These individuals did not arrive in New Mexico until February 1946, after processing captured documents and other material at Aberdeen Proving Grounds in Maryland. Although the German rocket experts possessed a nebulous official status, their role in the American rocket program was clearly defined. Von Braun and his colleagues were to serve as consultants to American industry and research institutions involved in rocket studies. They were to assist in the assembly and launching of V-2 rockets while conducting studies and proposing new projects. Advancing their own research agenda was not an option.[15]

Beginning in early 1946, White Sands served as the focus of the American missile program. Over the next six years crews launched sixty-four V-2s and a large number of smaller devices. Scientists and engineers from nearly every discipline associated with rocket theory visited White Sands at one time or another for launches or consultation. Exchanges of ideas and proposals led to new government and industrial rocket programs despite relatively modest financial support for White Sands activity. A V-2 Panel, established early in the program, coordinated V-2 research and included representatives of such agencies as the Naval Research Laboratory, General Electric (the army's prime contractor for rocket launches), the Johns Hopkins University Applied Physics Laboratory, and such universities as Harvard, Michigan, and Princeton. Although the first launch season, which began in April 1946 and lasted through the end of the year, returned only meager scientific results, crews and observers gained valuable experience in terms of launch activity and the retrieval of rockets.[16]

The complex nature of rocket launches provided ample opportunity for the participation of a wide variety of scientific and technical experts, some of whom were part of the Southwest's scientific establishment. Among the most encouraging results of the Applied Physics Laboratory V-2 flight of 30 July 1946, for example, was the completely successful warhead separation. Modifications of the separation system, which had caused various difficulties over the preceding few months, were developed by White Sands technical staff assisted by explosive experts brought in from New Mexico School of Mines at Socorro. Tracking V-2 flights had become a serious problem by the fall of 1946, leading White Sands officials to negotiate a contract with New Mexico College in Las Cruces for range personnel and preliminary data analysis. A member of the White Sands staff who soon coordinated the tracking program also had a significant southwestern connection. Clyde Tombaugh, who had discovered the planet Pluto from the Lowell Observatory in 1930, joined the Ballistics Research Laboratory at White Sands in 1946 and developed instruments and techniques to track rocket flights. By the end of the decade he was directing some eighty observers at more than fifty sites in southern New Mexico. Tombaugh would later join the faculty at New Mexico State University and establish its planetary astronomy program.[17]

By the time von Braun and his colleagues left White Sands for Huntsville, Alabama, in 1950, southern New Mexico had contributed significantly to the early research into rocket design and applications. Noted scientists James Van Allen and John A. Wheeler used V-2 launches to study cosmic rays at extreme altitudes, while others pursued solar research topics using rocket-borne spectrographic equipment. White Sands was the chief launch site for sounding rockets such as the WAC Corporal developed by the Jet Propulsion Laboratory and the Aerobee designed by the Applied Physics Laboratory. Other technical developments during the late 1940s included the initial launches of multi-stage devices and the early flight tests of the Viking rocket designed to replace the V-2. A few flights also contributed to biological research. Corn seeds and fruit flies were launched on selected high-altitude missions to analyze the genetic impact of cosmic rays. In 1948 and 1949, Project Albert launched two monkeys on V-2 flights to determine their response to high acceleration. In each flight the monkey survived the 5.5 G acceleration during liftoff and had no difficulties during the flight. The monkeys failed to survive their journeys, however, when parachutes failed to deploy.[18]

By the end of the 1940s the scientific installations at Los Alamos, Sandia, and White Sands appeared to symbolize the science of the American Southwest. Associated with military projects and federal funding, these facilities represented the "big science" that increasingly characterized American efforts in the postwar world. After decades of colonial status the region's scientists now participated in research that defined the nation's scientific outlook.

An exclusive focus on the large installations of the Southwest is, however, a limited and inaccurate perspective on the region's science. The new science of Los Alamos and White Sands redefined southwestern science but did not replace the scholarly pursuits that had characterized the region before the war. Topics that had interested scientists in earlier decades continued to offer opportunities for important discoveries and analyses. Although they often pursued their research in the new fashions of postwar American science, these scientists continued to expand knowledge in traditional areas and contributed to the dramatic growth of science in the American Southwest.

Federal agencies played a broader role in the region during World War II than that associated with the Manhattan Project. The Bureau of Mines and the U.S. Geological Survey, well established in the Southwest before the war, participated actively in the wartime expansion of science. Both agencies expanded their mineral exploration activity during the war, locating such strategically valuable minerals as manganese and molybdenum in Arizona, as well as new iron ore deposits throughout the region.[19] The continued interest in strategic minerals during the early years of the Cold War further expanded federal activity. The number of Geological Survey employees listed in *American Men of Science*, for example, increased from five in 1938 to eighteen in 1949. Employees of the Agriculture and Interior Departments played an important role in southwestern science as well, continuing research activity in crop production, forestry, and wildlife management.

Long identified with the region's scientific pursuits, archaeology remained an important activity during and after World War II. Archaeologists had little role to play in war-related research, but excavations and analyses continued throughout the 1940s. Among the more important analytical work was that of Harold S. Colton in Flagstaff on the Sinagua culture. By 1946, he had defined this culture as an amalgam of Hohokam, Anasazi, and other groups who moved into the Flagstaff area over time. Tracing various settlement phases, Colton described the development of this culture and its antecedents from circa 600 to circa 1450. University of Arizona anthropologist Emil Haury completed an important excavation during the war years at Ventana Cave west of Tucson on the Papago Reservation. With funds from the Indian Division of the Civilian Conservation Corps, Haury revealed nearly 11,000 years of cultural history and unearthed valuable evidence that the Hohokam culture developed from an earlier indigenous culture.[20]

Haury also pursued his investigations concerning the Mogollon culture during World War II, excavating sites in the Forestdale Valley as time and resources permitted. Seeking confirmation of his insight that the Mogollon culture was distinct from the Anasazi and Hohokam, Haury surveyed the eastern San Carlos Indian Reservation during the summer of 1945 to plan postwar fieldwork. The Point of Pines area proved especially attractive and became the site of major excavations and an important field school the following summer. Research and archaeological training continued at Point of Pines until 1960, by which time the unique and distinct nature of the Mogollon culture had largely been confirmed. Equally important, the field school served as the training ground for many of the region's leading archaeologists.[21]

The end of the war not only allowed the expansion of fieldwork in the Southwest but also witnessed important institutional developments. Annual archaeological meetings resumed in Santa Fe in August of 1946 at the Laboratory of Anthropology. Unlike those before the war, this conference was attended primarily by scholars from southwestern institutions. The universities of Arizona and New Mexico were well represented, as were the Museum of Northern Arizona and the Museum of New Mexico. Federal archaeologists and anthropologists came from the National Park Service, the U.S. Indian Service, and various national monuments. Another important institutional development took place during 1947 through the merger of the Laboratory of Anthropology, the School of American Research, and the Museum of New Mexico. The resulting state museum complex in Santa Fe played an important role in coordinating the expansion of archaeological endeavors in the Southwest. A similar concentration of resources emerged in Flagstaff during the late 1940s, as the Museum of Northern Arizona expanded its activities and facilities. By 1950 the museum established a separate research center nearby which included space for geological, botanical, and archaeological research, as well as living quarters for visiting researchers. The following year the Arizona State Museum in Tucson expanded its holdings by accepting the collections of the recently dissolved Gila Pueblo Archaeological Foundation.[22] With a solid institutional base and a growing number of scholars in the region (thirty-five individuals were listed in the 1949 edition of *American Men of Science*), southwestern archaeology was poised to significantly expand its endeavors.

The institutional base for astronomy in the Southwest was already well established before the war. Although the Steward and Lowell observatories remained moderately active, the McDonald Observatory in west Texas proved to be the center of southwestern astronomy during the 1940s. War demands depleted the observatory support staff, while many astronomers from the Yerkes/McDonald facilities were involved in wartime work for various agencies. Despite its reduced staff the McDonald Observatory remained active in astronomical work. Improvements to the equipment were completed during the war and several astronomers continued to make observations. Senior astronomers Otto Struve (Yerkes director) and George van Biesbrock, ineligible for military service, made heavy use of the large reflector on Mount Locke and coordinated the publication of the *McDonald Observatory Contributions*. On occasion, McDonald astronomers secured leave from their wartime assignments for limited research visits to the Texas facility. Few achieved more in this area than Gerard Kuiper in late 1943 and early 1944. Returning from the Harvard Radio Research Laboratory where he had worked on radar countermeasures, Kuiper concentrated on spectroscopic studies of the atmospheres of the major planets and satellites in the solar system. His observations revealed that Titan, the largest moon of Saturn, possessed an atmosphere of methane. This discovery represented the first identification of an atmosphere on a satellite and played an important role in redirecting Kuiper's research toward solar system astronomy.[23]

Following the war the McDonald Observatory attracted many leading astronomers as staff members or visiting astronomers. Continued improvement to the observatory's equipment further reinforced its status as one of the leading observatories in the world. Although McDonald astronomers pursued stellar and nebular research and occasionally participated in V-2 rocket flights at nearby White Sands Proving Ground, the Mount Locke facility increasingly focused on solar system astronomy after the war. Kuiper obtained U.S. Navy funding to pursue various topics. His observations of the major planets resulted in the discovery of the fifth satellite of Uranus (Miranda) in 1948 and the second moon of Neptune (Nereid) the following year. His infrared studies of planetary atmospheres allowed him to identify carbon dioxide in the Martian atmosphere and he also began an extensive study of the lunar surface using photographic collections from McDonald and other observatories.[24]

The Southwest contributed to the advance of astronomy in other ways after World War II. Harvard astronomer Fred Lawrence Whipple received $60,000 from the Naval Ordnance Laboratory for a series of new and fast meteor cameras in early 1946. A large sum for an astronomy project, the funds would support Whipple's project to study meteors to reveal characteristics of the upper atmosphere as well as information about the meteors themselves. While waiting for the completion of these new cameras Whipple conducted a careful survey of potential sites and chose the Las Cruces area for its high percentage of clear nights. He moved the existing meteor cameras from the Harvard College Observatory site in Massachusetts during the summer of 1947, employing this equipment for the next four years. The navy continued to fund Whipple's meteor research until 1954, after which the air force took over the project. Knowledge of the upper atmosphere was of increasing importance to the military during the space age, but also allowed astronomers such as Whipple to gain valuable knowledge for their discipline.[25]

The Lowell Observatory in Flagstaff emerged from World War II in a less than favorable position. The facility had no university affiliation and had not participated in wartime contract research. Finally, in 1947, scientists from the U.S. Weather Bureau approached the observatory to participate in a research project. The bureau proposed the Project on Planetary Atmospheres, an interdisciplinary study to gain insight into terrestrial atmosphere circulation. The Lowell Observatory was chosen for this project primarily because of its extensive photographic archive of planetary images. Funded by the Weather Bureau from 1947 to 1950, the project was taken over by the Geophysics Directorate of the Air Force Cambridge Research Center, responsible for upper atmosphere research for the U.S. Air Force. Although the Lowell Observatory continued to participate in this program for several years, the air force soon narrowed the Flagstaff research to solar constant studies. Nonetheless, the planetary atmospheres project indicated the continued value to astronomy of the Lowell Observatory and marked an important step in the postwar return of the facility as a major participant in American astronomy.[26]

Although southwestern scientists enjoyed expanded employment opportunities from both new and existing institutions, they continued to benefit greatly from the region's centers of higher education. These schools remained an important part of the Southwest's scientific foundation during the 1940s. As was the case throughout the United States, the

region's colleges and universities witnessed a dramatic increase in enrollment following World War II. Veterans took advantage of the GI Bill to
pursue college degrees, while younger students increasingly viewed an
undergraduate education as essential to later success. The University of
Arizona, for example, grew from some 3,000 students before the war to
more than 5,000 by the end of the decade. The University of New Mexico
grew by nearly 150 percent in this same period to 4,500, while Arizona
State Teachers College in Tempe (later renamed Arizona State University) nearly tripled in enrollment to more than 4,300.[27] Although faculty
grew much more slowly than enrollment, the number of scientists listed
in *American Men of Science* who held faculty positions showed a noticeable
increase.

An intriguing example of the region's educational establishment and
the growing importance of science in higher education was New Mexico School of Mines. The Socorro school nearly disappeared during World
War II as the student population all but vanished and the campus was
taken over by the military for educational programs. Uncertain of the
school's future, the Board of Regents authorized only a partial faculty of
four members for the 1944–45 academic year, a decision seemingly justified by the fall enrollment of fifteen students. As veterans began to return to the classroom the following fall enrollment increased to sixty-one
students, many of whom were especially interested in the engineering curriculum. The postwar focus of the school soon changed to broader scientific topics. E. J. Workman became president in 1946, moving from the University of New Mexico where he had been physics department chairman
and director of a major university research program. This latter program
was transferred to Socorro as the Research and Development Division
and soon began conducting contract research for private and government
agencies. The graduate program was established in 1946 as well, offering
a master's degree in earth sciences. Enrollment grew rapidly, exceeding
200 by the fall of 1947. The growing importance of the Socorro school as
a science center was clearly shown by the increase in the number of faculty members listed in *American Men of Science*. From five members in the
1938 edition, the number had grown to eighteen in 1949. The expansion
of the school's focus was also recognized in 1951 when the Board of Regents changed the name of the school to New Mexico Institute of Mining
and Technology.[28]

Table 6.1 Principal Employers of Southwestern Scientists, 1938 and 1949

| | *American Men of Science* Volume | | | |
| | 1938 | | 1949 | |
Institution	N	Percent	N	Percent
Los Alamos Scientific Laboratory	—	—	119	24
University of Arizona	61	28	76	15
University of New Mexico	23	11	35	7
U.S. Department of Agriculture	23	11	28	6
U.S. Department of the Interior	9	4	33	7
New Mexico College (Las Cruces)	16	7	31	6
New Mexico School of Mines	5	2	18	4
Texas School of Mines	4	2	10	2
Arizona State Teachers College	4	2	10	2
Sandia Laboratory	—	—	7	1

An analysis of the individuals listed in *American Men of Science* reveals the impact of World War II on the scientific community of the American Southwest. The institutional concentration of these scientists changed dramatically with the development of the Los Alamos and Sandia facilities (table 6.1). Although the major universities of the region remained important as employers of scientists, they no longer dominated the region. The postwar presence of the atomic research centers had a seemingly profound impact on the disciplinary distribution of the region's scientists. Physicists emerged as the new dominant group after the war, but chemists, engineers, and mathematicians also achieved new status. At Los Alamos, for example, 95 percent of those scientists listed in *American Men of Science* were chemists, physicists, or engineers; at Sandia Lab, the disciplines represented were physics, engineering, and mathematics. The broader disciplinary landscape of the postwar Southwest reflected the new science of the time (table 6.2). Chemists more than doubled their share of the population, while the proportion of physicists more than tripled. Those involved in biology and agriculture witnessed a decline in

Table 6.2 Specialties of Southwestern Scientists, 1938 and 1949

| | American Men of Science Volume | | | |
| | 1938 | | 1949 | |
Discipline	N	Percent	N	Percent
Chemistry	16	7	87	17
Agriculture	36	17	60	12
Biology	43	20	53	11
Engineering	25	12	68	14
Physics	8	4	73	15
Geology	21	10	38	8
Anthropology and archaeology	15	7	35	7
Mathematics	19	9	27	5
Other	32	15	61	12

their share of the scientific population, despite an increase in the actual numbers involved. These and other disciplines which had played major roles before the war, however, remained an important part of the scientific landscape. The number of archaeologists and anthropologists listed in *American Men of Science*, for example, more than doubled during the 1940s.

The dramatic growth of the region's scientific community had a notable effect on the distribution of educational backgrounds. At the undergraduate level (table 6.3), the University of Arizona continued to be the leading producer of the region's scientists, although its share of the population declined somewhat. The number of University of New Mexico undergraduates increased, revealing the Albuquerque school's emergence as an important regional educational center. Major universities such as Chicago, California, Michigan, and Harvard continued to be important sources for southwestern scientists at the undergraduate level. At the doctoral level (table 6.4), changes in distribution appear to have been precipitated by the new atomic research focus. The University of Chicago remained the leading producer of science doctorates for the region, combining its traditional

Table 6.3 Leading Undergraduate Programs of Southwestern Scientists, 1938 and 1949

| | *American Men of Science* Volume | | | |
| | 1938 | | 1949 | |
Institution	N	Percent	N	Percent
University of Arizona	17	8	30	6
University of Chicago	7	3	19	4
University of California	7	3	14	3
Ohio State University	2	1	15	3
University of Michigan	6	3	8	2
University of New Mexico	1	—	13	3
Harvard University	6	3	8	2

strengths in science with its important connections with the Manhattan Project. The University of Wisconsin, on the other hand, no longer claimed as dominant a position as it had before World War II. The Madison school had no strong program in atomic science and thus failed to supply large numbers of new scientists for the region.

The scientific community of the American Southwest grew dramatically between the late 1930s and the late 1940s. Much of that growth was a direct result of the establishment of Los Alamos during World War II and its growth after the war. Yet the presence of Los Alamos should not obscure the significant continuity that characterized the region's science at mid-century. The traditional focus of southwestern science, emphasizing natural history, applied science, and higher education, guided large numbers of scientists both before and after the war. This continuity does not, to be sure, indicate that prewar and postwar science was identical. By the late 1940s, the Southwest appears to have witnessed the emergence of two separate scientific communities. The first, continuing the tradition of prewar years, could be found in the region's universities, museums, and government agencies, teaching and conducting research in such fields as

Table 6.4 Leading Doctoral Programs of Southwestern Scientists, 1938 and 1949

| Institution | *American Men of Science* Volume | | | |
| | 1938 | | 1949 | |
	N	Percent	*N*	Percent
University of Chicago	18	16	32	11
University of Wisconsin	12	11	13	4
University of California	10	9	18	6
Harvard University	7	6	15	5
University of Illinois	3	3	19	7
Ohio State University	5	4	15	5

anthropology, agriculture, and biology. The second community, with access to sophisticated equipment, large budgets, and research teams that characterized "big science" could be found in laboratories in Los Alamos, Albuquerque, and Tucson, both on and off the university campuses in the latter two cities. These two communities provided the base for the continued dramatic growth in the region's science during the second half of the twentieth century.

The impact of World War II on science in the Southwest, therefore, involved the creation of a new scientific community to complement one that had steadily grown throughout the first four decades of the twentieth century. Although a crucial stage in the development of the region's scientific establishment, the war should not be seen as the primary factor in the creation of that establishment. Rather, World War II grafted a new variety of science onto the healthy trunk that had been growing for decades. The hybrid produced defined science in the American Southwest of the late twentieth century.

Part III The Foundations of Modern Science

Introduction

During the second half of the twentieth century the Southwest emerged as a full participant in the scientific developments that transformed American life. Although the Southwest's advances were based on earlier endeavors, their appearance in final form by the end of the century signified an important change in the region's intellectual status. No longer could a distinction clearly be made between "southwestern science" and that of other regions; a more accurate description would now be simply "science in the Southwest."

Demographic and political changes in the Southwest paralleled the emergence of the region's science. As a key actor in the Sun Belt phenomenon of the post–World War II period, the region's population increased dramatically from slightly more than 1.5 million in 1950 to nearly seven million by the end of the century. The population growth brought various benefits to the region, including a generally healthy economy and growing political influence. Individuals such as Arizona Senator Barry Goldwater attracted national attention throughout the period, but other figures played a more specific role in the Southwest's scientific endeavors. During the 1960s and 1970s, brothers Stewart and Morris Udall of Arizona addressed environmental concerns through their service in the Interior Department and the House of Representatives. New Mexico Senator Clinton Anderson chaired the Senate Space Committee and the Joint Atomic Energy Committee, both of which oversaw important developments in the Southwest's scientific infrastructure. During the 1990s, two southwesterners held important cabinet positions. Former Arizona Governor Bruce Babbitt served as secretary of the interior, while New Mexico's Bill Richardson directed the Energy Department.

Similarly important for the region's scientific growth was a dramatic increase in the size and scope of colleges and universities. The University of Arizona, long the leading educational institution in the region, witnessed a doubling of its enrollment each decade between 1950 and 1970, while Arizona State University skyrocketed from fewer than 5,000 students to more than 30,000 during the same period. As the region's schools increased in student and faculty populations, science became an increasingly visible aspect of their mission. External funding from private, federal, and state sources enabled universities to build new laboratories and hire scientists to conduct research projects in virtually all disciplines.

Science in the Southwest displayed great diversity during the last half of the twentieth century. The region continued to serve as a natural laboratory providing valuable data and insights. Archaeology and anthropology remained of great importance, as indicated by the presence of nationally recognized programs at the University of Arizona and the University of New Mexico. Hohokam irrigation canals in the Phoenix area were excavated during the 1970s under the auspices of the Arizona State Museum and the National Park Service. Despite widespread development in the area, much of interest was still recoverable through traditional techniques and archival studies of aerial photographs. Environmental topics were investigated in the region throughout the postwar period. The Irving Langmuir Laboratory for Atmospheric Research was established in New Mexico in 1961. Funded with a $500,000 grant from the National Science Foundation, the facility was administered by the New Mexico Institute of Mining and Technology. Biologists from the University of Arizona and New Mexico State University studied the Jornado Experimental Range in southern New Mexico in an effort to better understand the impact of climate and long-term grazing on desertification in the region.[1]

The Laboratory of Tree-Ring Research continued to be the focal point for dendrochronological studies throughout the last half of the century. In the 1950s the laboratory pioneered the analysis of the ancient bristle-cone pine in California, eventually developing a complete chronology back to 6700 B.C. During the 1960s this chronology played an important role in the correction of the radiocarbon dating method developed earlier. The laboratory worked with various radiocarbon facilities to develop a calibration curve to correct the dates that had been shown to be in error. Although archaeological research remained an important part of the laboratory's efforts, dendrochronologists in Tucson increasingly used the tree-ring record to reconstruct past climates, charting seasonal and long-term fluctuations. By the end of the century they were also employing chemical analysis of various isotopes to study tree growth characteristics in past climates.[2]

The region's economic base continued to benefit from scientific activity. Although commercial agriculture now represented only one aspect of the Southwest's economy, its continued well-being increasingly depended on research conducted by scientists at the region's universities and agricultural experiment stations. An especially dramatic example of this activity was the role played by scientists in the New Mexico chile industry beginning in the 1940s. Nutritionist Edith M. Lantz joined the faculty at New Mexico College of Agriculture and Mechanic Arts in 1935 and began investigating regional foods, including chile. She initiated a study of the carotene and ascorbic acid content of this dietary staple, publishing a report in 1943 that emphasized that all varieties tested were rich in vitamin C and nearly all were rich in carotene. Two years later Lantz conducted a more focused study comparing García's "New Mexico No. 9" chile with Anaheim peppers and reported that the New Mexico product was richer in vitamin C.

Lantz focused on preservation in a 1946 study of the effects of canning and drying on the nutritional content of chile. Green chile retained practically all its carotene during canning and lost no more than one third of its ascorbic acid. During storage, a gradual but slight loss of both substances could be measured. Green chile dried by artificial heat produced a good product with significant amounts of vitamin C and carotene, although canned chile remained higher in both. Preserved red chile was much less valuable as a source of ascorbic acid, losing most of its vitamin C while maintaining a high level of carotene. Lantz also examined commercial chile products for her study, finding that both ascorbic acid and carotene values varied significantly.[3]

Lantz's research provided important information about a food of regional significance, but chile had not yet established itself as a commercial product at the national level. Following the successful introduction of "New Mexico No. 9" the state's chile industry had expanded slowly during the 1920s and 1930s. Although García's chile was superior to native varieties in terms of pungency and suitability for canning, the industry recognized that an even milder chile with a better shape for processing was necessary to establish New Mexico chile in the national market.

Horticulturist Roy Harper of New Mexico College in Las Cruces began attempts to develop a more suitable chile in the late 1940s. Crossbreeding chiles of unknown genetic origin obtained from local commercial fields, Harper developed several varieties of which the strain later called "New Mexico No. 6" proved most desirable. Not only was it milder than No. 9 (and thus more marketable to consumers outside the Southwest), it yielded more fruit in pods that were easier to process. By 1953, No. 6 was well established as a dual-purpose variety, providing green chile for canning and red pods for dry powder. Harper soon developed an even milder version of this chile, releasing "New Mexico 6-4" in 1959. Because of its flavor and greater suitability for canning, this variety soon dominated the market and was largely responsible for the rapid growth of the industry in the 1970s.

Harper and his successor, Roy Nakayama (the son of a Japanese farmer who settled in the Mesilla Valley in 1918), developed numerous chile varieties into the 1980s, after which the New Mexico State University program continued similar research. New breeding techniques based on tissue culture and genetic analysis were quite different from the traditional methods of cross-pollination and field testing, although the latter remained an important part of the process to determine commercial viability. A further refinement of variety 6-4 led to the 1990 release of "NuMex Joe E. Parker," a cultivar with greater yield and less variability, attributes that were of great interest to commercial growers.[4]

The chile varieties developed by the horticulturists in Las Cruces created the base for the growth of the New Mexico chile industry into a significant national economic force. Reflecting the industry's regional focus, chile acreage in New Mexico during the 1950s had remained stable at slightly more than a thousand acres. The development and acceptance of Harper's new varieties led to a fourfold increase by the early 1970s, after which chile cultivation expanded rapidly. By the end of the decade, New Mexico growers cultivated some 15,000 acres, a figure that more than doubled by the early 1990s. Clearly a major international business, the state's chile industry remained an important part of the New Mexico economy at the opening of the twenty-first century.[5]

Research and development in technology became increasingly important in the postwar Southwest as well. Weapons research at Los Alamos and Sandia laboratories remained central in this context, with the Los Alamos budget in the mid-1980s approaching half a billion dollars. Aerospace and electronics facilities appeared in most urban areas, building on the foundation laid during World War II. During the 1950s the expansion of both Motorola and AiResearch in the Phoenix area established the Valley of the Sun as an important "high tech" region. Cold War research benefited these and other companies; by the mid-1960s, AiResearch was the second largest employer in Arizona. In Tucson the Hughes Aircraft plant produced missiles for the U.S. Air Force. Prospering in times of high defense spending, the Tucson plant employed more than nine thousand workers in the mid-1980s. Computer giant Intel Corporation built a huge plant in Rio Rancho, a northwest suburb of Albuquerque, and soon grew into one of northern New Mexico's major economic forces. By the late 1990s, Intel employed more than six thousand workers and contributed an annual payroll of $260 million. Intel plants in Phoenix and Chandler, Arizona, made similar contributions to that state's economy.[6]

Also building on wartime and immediate postwar events, the Southwest witnessed a major expansion of space-related research and development. White Sands Missile Range continued to be a major center for this activity even after the transfer of the von Braun team to Alabama. By the late 1950s more than 6,700 people worked at the facility, which had a budget of nearly $160 million. Holloman Air Force Base, near Alamogordo, became a center for aviation medicine. Rocket-powered sled experiments were crucial to the development of safe ejection techniques for high-speed aircraft. As the space program progressed during the late 1950s and 1960s, southwestern facilities became increasingly important. The Holloman Aeromedical Laboratory coordinated the rigorous training and testing of the chimpanzees used in the early Mercury flights, providing valuable information for the later manned missions. Apollo astronauts were involved in training activities throughout the Southwest. They visited Kitt Peak National Observatory near Tucson to study telescopic images of lunar craters and participated in geology field training exercises at the Grand Canyon, in the Big Bend country of Texas, at Arizona's Sunset Crater, and near Cimarron, New Mexico. Sunset Crater was especially valuable in training astronauts to walk across volcanic surfaces with minimal risk of damage to spacesuits.[7]

Among the most dramatic advances in the region's science were those in astronomy. The Southwest had long been recognized as a superior site for visual astronomy, but the institutional base had only slowly established itself before the late 1950s. In March of 1958, after four years of careful study, southern Arizona's Kitt Peak was announced as the site for the national observatory. The 2,400-acre site leased from the Tohono O'odham reservation southwest of Tucson soon held several major instruments, including an 84-inch reflector completed in 1961. A large solar telescope funded with $3.5 million from the National Science Foundation began operating in mid-1962. The following year officials began plans for a major telescope that would be the centerpiece of the national observatory. This instrument became the 158-inch Mayall Telescope completed in 1973 at a cost of $10 million.[8]

The establishment of Kitt Peak National Observatory led to significant expansion of the region's astronomy. As the American space program grew in the late 1950s and early 1960s, NASA discovered that planetary astronomy was plagued by various problems. The agency quickly began providing money to support ground-based astronomy, much of which found its way to the Southwest. Grants and contracts were awarded to the Lowell Observatory, the universities of Arizona and New Mexico, and New Mexico State University for various programs and projects. The first major planetary telescope funded by NASA was the 61-inch reflector constructed between 1962 and 1965 by the Lunar and Planetary Laboratory of the University of Arizona. Built in the Catalina Mountains north of Tucson, this instrument became the first of many telescopes constructed in the area. NASA provided money for other projects as well, including $5 million for the 107-inch reflector at McDonald Observatory in Texas (constructed between 1963 and 1967) and more than a million dollars for a separate Space Sciences Building on the University of Arizona campus.[9]

The University of Arizona benefited greatly from the expansion of the region's astronomy. The National Science Foundation provided Steward Observatory with $1.4 million in 1965, most of which financed construction of a 90-inch reflector for the facility's site on Kitt Peak. During the 1970s the university joined forces with the Smithsonian Astrophysical Observatory to design and build an entirely new type of telescope. The Multiple-Mirror Telescope arranged six 71-inch mirrors in a computer-controlled mounting to provide the equivalent light-gathering power of a 180-inch instrument. The university's Optical Sciences Center had the six mirrors as a result of a cancelled military satellite project, and the Smithsonian had an excellent site south of Tucson on Mount Hopkins. Despite the new and challenging technological demands the instrument was completed in 1978 and confirmed the value of this new telescope design.[10]

Optical astronomy was only one aspect of the study of the heavens. Radio astronomy was increasingly of interest during the 1960s and led to another important advance in southwestern science. Astronomers at the National Radio Astronomy Observatory realized that several radio telescopes working together would provide the equivalent of a much larger instrument. The Very Large Array design evolved during the late 1960s and received congressional authorization in 1972. The thirty-seven dishes, each eighty-two feet in diameter and weighing 230 tons, would be mounted on railroad tracks so they could be moved as needed to provide the required observing characteristics. Designed to give an equivalent resolving power of a single telescope sixteen miles in diameter, the Very Large Array was located on the Plains of San Augustin, fifty-two miles west of Socorro, New Mexico. The array was completed in 1981 at a cost of some $78 million and immediately began returning valuable data. The administrative and scientific headquarters were moved to a large building on the campus of New Mexico Tech following a $3 million appropriation from the State of New Mexico.[11]

As the 1980s merged into the 1990s, the role and status of science in the American Southwest became increasingly varied. Advances in astronomy continued to be dramatic. Roger Angel and his colleagues at the Steward Observatory developed a new method of casting telescope mirrors that would lessen the weight of these devices and reduce the time and money required to finish them. This "spin-casting" technique employed a rotating oven and quickly proved its worth, convincing the National Science Foundation to fund Angel's efforts. By the early 1990s several mirrors had been successfully cast, including three of 138-inch diameter. One of these went to the Apache Point Observatory near Cloudcroft, New Mexico, operated by New Mexico State University for a consortium of universities.[12]

The region's scientific eminence led to efforts at further expansion. With a well-established institutional base, the Southwest saw no reason not to campaign for the major science project of the 1980s: the Superconducting Super Collider (SSC). Recommended by the Department of Energy in 1983 and endorsed by President Ronald Reagan in 1987, the SSC was a giant particle accelerator that would require a circular tunnel more than fifty miles in length. Although the collider would provide significant information concerning subatomic physics, much interest was also generated by the project's estimated cost of more than $4 billion, the prospect of 5,000 construction jobs, and a permanent employment level of some 2,500 people once the facility was completed. Among the forty-three proposals filed by twenty-five states seeking this project, five were located in the Southwest. The site thirty-five miles southwest of Phoenix survived the review process and was one of eight locations recommended for further study in early 1988. Land availability, favorable geology, and a large pool of technical workers in the Phoenix area represented important advantages, although operating costs were likely to be slightly higher than at other sites. The Department of Energy ultimately chose a site south of Dallas, Texas, in November 1988, but political and scientific support for the SSC steadily eroded over the next five years. Faced with revised cost estimates of $11 billion, Congress voted to cancel the project in October 1993.[13]

That the Southwest sought the SSC suggested that the region had become a significant player in the nation's science. Yet the region's status failed to insulate it from anti-science sentiment. In 1996, for example, the New Mexico Board of Education approved revised standards for public school science that pointedly deleted references to evolution and were applauded by those who accepted "creation-science" as a viable alternative.[14]

The protracted struggle to build an observatory on Mount Graham in southeastern Arizona was a particularly dramatic example of the ambiguous status of science in the Southwest. Announced in the mid-1980s, the proposed observatory would be coordinated by the University of Arizona but would include several other observatories from around the world. Working with the U.S. Forest Service and other agencies, university astronomers designed their plans to minimize the impact of the observatory on the habitat of the Mount Graham red squirrel, recently placed on the endangered list. Although the observatory site would include no more than twenty-four acres of the 11,000-acre red squirrel habitat, a small but vocal group of self-styled environmentalists campaigned against the observatory in public and in court for more than a decade. The Apache Survival Coalition filed a suit claiming that the development of the observatory would desecrate ancient religious sites. Congressional intervention was eventually necessary to end the ordeal. The federal budget passed in 1996 included a specific authorization of the new site on Mount Graham as consistent with earlier legislation.[15]

Other difficulties faced southwestern science during the 1990s, most of which involved financial concerns. The end of the Cold War redirected efforts at Los Alamos and Sandia from weapons research toward other activities. Los Alamos National Laboratory included a large number of astrophysicists and plasma physicists who were engaged in fundamental research. Other Los Alamos scientists pursued fusion research, global climate modeling, environmental monitoring, and research concerning the disposal of weapons-grade plutonium. Sandia National Laboratory pursued fusion energy research and environmental projects.[16] Concerns about the federal budget frequently led to cuts in various science programs in the Southwest. In the late 1980s and again in the mid-1990s, funding shortfalls for ground-based astronomy led the National Science Foundation to close certain facilities on Kitt Peak. Maintenance budgets were also pared to the bone. The Very Large Array suffered especially from this situation, leading a science writer to describe the facility as "Radio Astronomy's Crumbling Showpiece."[17] Although the Southwest was certainly not the only region suffering from such difficulties, financial problems and their solutions appeared to compromise the previous fifty years of dramatic growth in science.

The Southwest's emergence as a full partner in the scientific pursuit by the end of the twentieth century was a product of earlier efforts. As displayed by the growth of astronomy in southern Arizona before the establishment of Kitt Peak National Observatory, the region's scientists were able to make contributions despite the lack of financial and other support. These contributions did much to establish the Southwest as a region with potential for scientific advances. As was the case elsewhere in the nation, however, science did not exist in a vacuum. The impact of science on the general population remained a constant in the twentieth century and frequently involved scientists in debates they would rather have avoided. Although not as well known as events elsewhere in the United States, Arizona's evolution controversy was another example of the region's place in the nation's scientific culture. The Southwest also participated in the major change in science in the postwar world, the emergence of "big science." One of the most dramatic examples of this phenomenon was the American space program. From the early 1960s through the end of the twentieth century, scientists from throughout the region participated fully in the various programs designed to explore space. By the opening years of the twenty-first century, the American Southwest had become a crucial component of the nation's scientific establishment.

Chapter 7 Astronomy in Southern Arizona, 1889–1963

The selection of Kitt Peak in 1958 as the national observatory site catapulted the region into a leading role and alerted American scientists to what those who had worked in the area already knew: the clear steady desert air of southern Arizona was superior for astronomy. Despite sparse population and inadequate institutional resources, southern Arizona had contributed to the growth of astronomy in the United States for more than a half century before the establishment of Kitt Peak National Observatory. Throughout this period the University of Arizona played a central role in laying the foundation for later developments.

Following its establishment in 1889, the territorial university represented little more than a hesitant attempt to provide an educational base for the region. The university stressed the economically important fields of agriculture and mining, although astronomy claimed a minor presence from the beginning. A four-inch refracting telescope was donated to the university in 1891 and served as a teaching tool when coursework in astronomy began five years later. The astronomy curriculum tended to focus on practical aspects such as surveying, but courses were only offered intermittently.[1]

The fortunes of astronomy at the University of Arizona began to change in the fall of 1906 with the arrival of A. E. Douglass as a member of the science faculty. Although he began the revitalization of astronomy by developing a two-semester survey course to begin in the fall of 1907, Douglass's recognition of the superior atmospheric qualities of southern Arizona convinced him that the region was an excellent site for a large observatory. Support for such a project, however, was nowhere to be found, as the territorial legislature viewed a large telescope as an unnecessary addition to the university's equipment. Taking advantage of his earlier connection with Harvard, Douglass arranged to borrow an eight-inch refractor in the spring of 1908. He convinced the University of Arizona to construct an observatory room on the top floor of the newly completed Science Hall and installed the Harvard telescope during March 1909. Douglass used the telescope for the next dozen years, observing planetary, cometary, and stellar phenomena. During January 1910, for example, he turned the borrowed instrument toward Comet *a*, carefully recording the comet's appearance. His account of these observations along with several of his drawings appeared in the March issue of *Popular Astronomy*.[2]

Douglass refused to abandon his goal of a major astronomical facility in southern Arizona and continued to explore prospects for funding. By the fall of 1914 he had collected letters of support from such leading astronomers as George Ellery Hale (Mount Wilson Observatory) and Edwin B. Frost (Yerkes Observatory), and had corresponded with the famous Pittsburgh lens-maker John A. Brashear and Company to obtain cost estimates for various telescopes. Douglass also sent letters to civic leaders in the state, many of whom he knew personally, asking for their help. If leading financial figures in the state would donate sufficient money as an "entering wedge," he argued, the legislature might be encouraged to appropriate additional funding for the observatory. Despite these endeavors the Tucson astronomer was unable to convince the legislature of the value of a large telescope.[3]

Douglass's campaign remained moribund until the summer of 1916, when he once again began sketching plans for a large facility. The reasons for Douglass's renewed activity became clear in mid-October when University of Arizona officials announced that an anonymous benefactor had given the university $60,000 for an observatory. As became known later, this gift came from Lavinia Steward, a resident of southern Arizona since 1898. She and her husband Henry B. Steward had moved to the area from

Joliet, Illinois, after retiring from the flour milling business. During the summer of 1916 Mrs. Steward told friends that she would like to do something for the University of Arizona in memory of her husband, who had died in 1902. An amateur astronomer herself, she proved very receptive to the suggestion that she endow an observatory.

Douglass's many years of planning now proved valuable. By the end of the year he had drafted a comprehensive plan for the Steward Observatory and arranged for the optical work to be done by Brashear. The mounting would be constructed by the Cleveland engineering firm of Warner and Swasey, who had done similar work for the Lick Observatory in California. Douglass surveyed various sites for the new telescope, finally deciding to erect the facility on the east end of the University of Arizona campus. Throughout these early efforts Douglass focused his attention on securing the largest telescope possible, even if compromises were necessary in other considerations.[4]

Douglass's plans for a rapid completion of the observatory, however, were doomed to failure. As World War I became a greater concern to the United States, American manufacturers were increasingly involved in war work. Astronomical equipment remained a low priority. Similarly frustrating was the lack of European sources for mirror blanks. Douglass had to arrange for American glass manufacturers, none of whom had experience casting and annealing large mirrors, to experiment with the thirty-six-inch mirror for the Steward telescope. Ultimately choosing the Spencer Lens Company of Buffalo, New York, for the project, Douglass waited until December 1921 for the firm to cast a successful disk for Brashear opticians to turn into a telescope mirror. The observatory building in Tucson had been complete for nearly a year and the mounting for six months; grinding and polishing the mirror was the last step. On 5 July 1922, Brashear workmen delivered the primary mirror and all optical accessories to the express company office in Pittsburgh. Five days later the shipment reached Tucson and was immediately delivered to the observatory. Douglass made the initial view through the instrument on the first clear day, 17 July, bringing the crescent Venus into focus as "first light" was achieved.[5]

The installation of the thirty-six-inch mirror in July 1922 brought the Steward Observatory into the mainstream of American astronomy. The size of the mirror and the clear desert air combined to establish an astro-

The Steward Observatory building on the University of Arizona campus as it appeared in 1981. (Photograph by the author)

nomical outpost with great potential. With the arduous task of completing the facility behind him, Douglass could direct his attention to the equally demanding problem of developing that potential.

The first major research project of the Steward Observatory, however, did not involve the new thirty-six-inch telescope. The solar eclipse of 1923 would be visible in the Southwest on 10 September and represented an important astronomical event Douglass hoped to observe. For the past five years he had collected information and equipment for a planned expedition to northern Mexico, the nearest site along the line of totality. By early August he had selected an observing site at Puerto Libertad, on the east coast of the Gulf of California, and had begun to assemble the eclipse equipment. The university provided a $500 budget for the expedition but this was to include all transportation and food as well as observing apparatus. From mirrors and lenses loaned by the U.S. Naval Observatory, the Carnegie Desert Botanical Laboratory in Tucson, and several area residents, Douglass and his assistants assembled four instruments to record photographically the progress of the eclipse. Leaving Tucson in early September, the eclipse expedition reached the Puerto Libertad site on the sixth and began the complex task of assembling and testing the equipment.

Working quickly during the morning of 10 September, the astronomers conducted further tests and made final adjustments. As the time for the eclipse neared, Douglass entered the lightproof shelter housing the plate holder of the largest instrument. Here he would carefully expose as many plates as possible during the course of the eclipse. The drama of a total solar eclipse proved an impediment to observation—the observatory crew who remained outside to inform Douglass of the beginning of the eclipse were so transfixed by the spectacle that some ten to twenty seconds passed before anyone remembered to give word to begin the photographic observations. The number of exposures taken of the eclipse was thus limited although the weather cooperated with clear skies and little wind.

After dismantling and packing the expedition's equipment, Douglass and three astronomy students left Puerto Libertad on Thursday, 13 September, reaching Tucson Friday evening. For the next two days they worked feverishly to develop and print the plates exposed during the eclipse. Carrying his negatives and one undeveloped plate, Douglass rushed to catch Sunday's midnight train to Los Angeles to present the results of his expedition to the meeting of the Astronomical Society of the Pacific. In Los Angeles a pleasant surprise greeted the Tucson astronomer. The California coast had been covered by clouds on the day of the eclipse, making photography impossible. Although a few marginal observations had been made in the eastern United States, the Steward Observatory expedition returned the only superior record of the 1923 solar eclipse.[6]

Although much of Douglass's time and energy were devoted to improving the physical and financial status of the Steward Observatory, research continued to be an important consideration. In the spring of 1924 Douglass began a program of planetary photography in preparation for the summer's Martian opposition. When the red planet made its periodic close approach to Earth in August and September, Douglass and his assistants were able to record a large number of drawings and photographs of the Martian surface. The photographs, made through various filters, supported the widely held view that the dark surface markings represented some form of vegetation. The "unmistakable" and "very evident" green color of these surface markings, plus their deepening color as the polar caps melted, seemed to provide evidence that the red planet was in important ways similar to Earth. The Steward Observatory continued to turn its large telescope toward Mars during the remainder of Douglass's tenure as director, confirming at each opposition the likelihood of vegetative life on the planet. This view represented astronomical orthodoxy until the unmanned space probes of the 1960s and 1970s.[7]

While beginning his observations of the 1924 Martian opposition, Douglass resumed his campaign to improve the Steward Observatory. To this end, he repeated his request for the appointment of an assistant in the astronomy department, to allow Douglass to concentrate on his many research projects and to add courses to the astronomy program. This significant step forward was taken in early 1925 with the appointment of Edwin F. Carpenter as Assistant Professor of Astronomy. With bachelor's and master's degrees from Harvard (where he had been elected to Phi Beta Kappa) and a recently completed doctorate from the University of California, Berkeley, Carpenter presented a valuable addition to the observatory staff. He quickly assumed the bulk of the class work, establishing himself as a popular and stimulating teacher who greatly expanded the department's course offerings during the next few years.[8]

Despite the absence of such needed equipment as a spectroscope and a student telescope, the Steward Observatory appeared to be making a concerted effort to realize its potential. The impact of inadequate support, however, continued to compromise research efforts, as shown by Carpenter's inability to publish quickly the results of earlier work. As a doctoral student he had taken nearly three dozen spectrograms of the well-known eclipsing variable U Cephei using the thirty-six-inch refractor at Lick Observatory. His observations indicated a highly eccentric orbit for the brighter component of this system, in contrast to the nearly circular orbit suggested by visual data. In order to reconcile these conflicting interpretations Carpenter hoped to make further observations at the Steward Observatory, but five years after joining the Arizona faculty he still had no access to a spectrograph. Although Carpenter eventually published an article on his work in the prestigious *Astrophysical Journal* his contribution remained suggestive rather than definitive.[9]

The value of the Steward Observatory's location and equipment was dramatically confirmed in November 1928 when the noted astronomer Edwin P. Hubble visited the facility. Recently elected president of the International Astronomical Union's Committee on Nebulae, Hubble was in Tucson to coordinate the observatory's participation in a cooperative photographic survey of bright nebulae. Under the auspices of the IAU the survey would include such telescopes as the Yerkes twenty-four-inch reflector in Wisconsin, the Mount Wilson sixty-inch and Lick Observatory thirty-six-inch reflectors in California, the forty-inch reflector at the Lowell Observatory in Flagstaff, and the Steward Observatory's large telescope. Despite Carpenter's increasingly demanding duties the young astronomer eagerly accepted Hubble's invitation to join the project.[10]

In addition to providing valuable data on nebulae visible south of Tucson (the area of the sky assigned by Hubble), Carpenter reported other findings to various professional organizations during the early 1930s. In June 1931, Carpenter presented two papers at the joint meeting of the Astronomical Society of the Pacific and the American Association for the Advancement of Science in Pasadena, California. His paper, "The Distribution of Color in Two Extra-Galactic Nebulae," provided important information concerning the different types of stars found in the nebulae M51 and M82. His other contribution, published in the *Publications* of the Pacific organization as "A Cluster of Extra-Galactic Nebulae in Cancer," detailed Carpenter's discovery of a cluster of some sixty nebulae in the western part of the constellation Cancer, a discovery made at the same time by Hubble at Mount Wilson. This discovery brought significant recognition to Carpenter and the Steward Observatory and was cited in *Transactions of the International Astronomical Union* and the *Astrophysical Journal*.[11]

Carpenter's most significant discovery from his nebular work, however, was announced at the September 1931 meeting of the American Astronomical Society at Perkins Observatory in Delaware, Ohio. Carpenter became the center of attention when he announced his discovery of an absorption layer near the plane of the Milky Way galaxy. Confirming the 1930 work of R. J. Trumpler of Lick Observatory, Carpenter's preliminary findings indicated a gradual diminution of nebular brightness as the galactic plane was approached, leading Carpenter to suggest that many estimated galactic distances were as much as ten times too large. Although his estimate of the effect of this absorbing layer of dust proved excessive, Carpenter's work provided important data for the refinement of astronomers' ideas concerning galactic dimensions. When the Steward Observatory completed its survey tasks in 1935, Carpenter's contributions had done much toward establishing the facility as a respected member of the astronomical community.[12]

Despite the contributions made and the reputation enjoyed by the Steward Observatory the facility remained in a precarious position. Not only did the observatory continue to lack adequate space and equipment because of the low levels of university funding, but the telescope itself was threatened by Tucson's growth. As the city and the university expanded, the accompanying light interfered with the long-exposure photography which characterized modern astronomy. By the spring of 1930 Douglass had completed a preliminary survey of nearby locations and had identified a promising new site for the observatory. A dozen miles east of Tucson, in an area that would later become part of the Saguaro National Monument, he located a low hill surrounded by rising ground for excellent protection against wind and lights. Douglass provided President H. L. Shantz with two detailed plans for establishing an observatory on this hill. The least expensive proposal called for the transfer of the Steward reflector to the new site and its replacement on campus by a ten- to twelve-inch refractor. This plan would cost the university some $50,000. A more ambitious proposal would retain the thirty-six-inch reflector on campus and erect a sixty-inch instrument on the site east of town. The $300,000 cost of this plan made it unlikely. Although Douglass and Carpenter continued to develop plans for the site throughout the 1930s and to encourage university officials to take steps toward securing the relocation of the Steward facility, financial considerations made the transfer impossible.[13]

Despite the observatory's inability to secure a more favorable site, astronomy at the University of Arizona had undergone an important change by the late 1930s. Douglass had increasingly focused his energies on the new science of dendrochronology, gradually moving away from active involvement in the astronomy program in Tucson. Before his retirement at the age of seventy in June 1937, Douglass had secured an additional permanent position in astronomy at the university, hiring Franklin E. Roach as Assistant Professor of Physics and Astronomy. Roach's undergraduate training at the University of Michigan was followed by graduate degrees at the University of Chicago (Ph.D. in astrophysics, 1934), where he served as an assistant at Yerkes Observatory. Roach's additional duties in physics instruction, however, limited the contribution he could make to the astronomy program. When Carpenter became director of the Steward Observatory in the summer of 1938, he thus faced many of the same difficulties his predecessor had faced since the early 1920s. Inadequate university financial support, limited space and equipment, and an increasingly unacceptable site raised serious questions about the future of the Steward Observatory.[14]

The late 1930s nonetheless provided hopeful signs for the realization of the observatory's potential. Despite his many duties in both astronomy and physics, Roach pursued various research topics in Tucson. He supervised the construction of a photometer by two graduate students in the astronomy department and employed this new device in the investigation of the minor planet Eros in February 1938. His photometric observations, using a technique that would become increasingly important during the years ahead, allowed him to describe the body as an ellipsoid with three unequal axes and to refine observations made by other astronomers earlier in the decade. During the opposition of this minor planet in the late spring of 1940, Roach made more than 300 exposures of Eros to confirm his earlier calculations of the size of the asteroid. Roach's discussion of his research appeared in two articles in *Astrophysical Journal*.[15]

The Steward Observatory also attracted the interest of Willem J. Luyten (University of Minnesota), the leading expert on white dwarf stars. Luyten had been locating and studying these objects since their initial discovery in 1920, examining photographic plates at the Harvard College Observatory and making observations with the Lick thirty-six-inch refractor. By 1939 he had completed this aspect of his survey and determined that additional white dwarfs could best be discovered through the analysis of stars with large proper motions. Luyten approached Carpenter for observing time on the Steward reflector and received an enthusiastic reply. Partially funded by a $500 grant from the National Academy of Sciences, Luyten began his work in Tucson during the 1940–41 academic year, recording the colors of more than 200 stars for later analysis. Similar observations were made at the Cordoba Observatory in Argentina under Luyten's direction. These small stars, approaching the end of their stellar evolution, continued to be Luyten's chief interest and remained an important aspect of the Steward research program for the next two decades.[16]

During World War II the astronomy program at the University of Arizona was characterized by temporary measures. Carpenter served as a naval reserve officer, teaching navigation at Cornell, while Roach moved to the California Institute of Technology to participate in activities for the Office of Scientific Research and Development. University officials asked Douglass to assume the acting directorship of the Steward Observatory and hired two astronomers as temporary faculty members. These two astronomers (Paul D. Jose and Joseph F. Foster) taught the basic astronomy courses as well as war-related classes such as navigation and "Elementary Military Astronomy" while assisting visiting astronomers who continued to use the Tucson facility for their research. Although Douglass and Jose investigated various locations in southern Arizona, efforts to secure an improved site for the Steward Observatory were unsuccessful. The wartime university budget could not be stretched to cover the expenses involved in moving the Steward reflector.[17]

The end of World War II and the return of Carpenter as director of the observatory had little immediate impact on southern Arizona astronomy. Constrained by an increasingly inadequate location and discouraged by a decade of subsistence funding, Carpenter expected no significant improvement. The beginning of the 1946–47 academic year, however, provided an encouraging confirmation of the potential for astronomy in southern Arizona. Frank Bradshaw Wood (Ph.D., Princeton, 1941), the last student of the eminent astronomer Henry Norris Russell, chose the Steward Observatory for his work as a National Research Council Fellow. A specialist in eclipsing binary stars, Wood was one of eighteen fellows appointed for the academic year and the first research fellow to choose the University of Arizona for his work. Light interference and polluted skies at the Tucson facility forced him to transfer his research activity to Lick Observatory in the summer of 1947, leading Carpenter to remind University of Arizona President Alfred Atkinson of the observatory's continuing difficulties. Financial support for equipment and staff remained at a low level and, although the university was in the midst of its largest building program ever, the observatory had not been included in the expansion. Although Carpenter persuaded Wood to accept an appointment at the University of Arizona in 1947, the young astronomer left Tucson three years later for a more advantageous position at the University of Pennsylvania.[18]

The problems facing the Steward Observatory failed to dampen the enthusiasm of researchers who continued to visit Tucson for visual and photometric observations. With funding from Sigma Xi, the National Academy of Sciences, the American Academy of Arts and Sciences, and other groups, Willem J. Luyten continued his white dwarf survey with the assistance of Carpenter and others. After a decade of work with the Steward Observatory, Luyten could report on significant progress by early 1950. He had visited the Tucson facility on twelve occasions between 1940 and 1949, but had also relied on Steward staff for additional observations. More than 600 hours of observing time had been devoted to a survey of some 2,000 stars, 40 of which Luyten identified as white dwarfs. The Minnesota astronomer also identified twenty-nine stars of this type from material furnished by the Cordoba Observatory and found four additional white dwarfs on plates from the Harvard and Van Vleck observatories. The seventy-four stars identified from this survey represented two-thirds of all white dwarfs known in 1950. Luyten continued his visits to Tucson through the late 1950s, although the deterioration of the observatory's atmospheric conditions minimized new discoveries. His articles in professional journals (frequently co-authored with Carpenter) cited the help of the observatory staff and provided the facility with valuable publicity.[19]

Carpenter and his staff continued to conduct their own research as well. Walter S. Fitch had joined the Steward staff in 1951 while still a doctoral student at the University of Chicago. He pursued his research on the variable star VZ Cancri using the Steward reflector to augment observations made in 1950 and 1951 at the Yerkes and McDonald observatories. By April 1954, Fitch had recorded more than 1,400 photometric observations of this star and had begun drafting a report of his work for *Astrophysical Journal*. This important article was published in May 1955 shortly before the award of Fitch's Ph.D.[20]

In addition to his research and administrative duties Carpenter became active in yet another project during the early 1950s. The many changes in astronomy during the previous decade and the growing awareness of the inferior sites of many American observatories led astronomers to consider new tactics in their efforts to explore the universe. Beginning in late 1951, Carpenter began corresponding with colleagues at Goethe Link Observatory (Indiana University) and Perkins Observatory (jointly operated by Ohio Wesleyan and Ohio State University). This correspondence resulted in a 1952 proposal to the National Science Foundation for a cooperative observatory in the American Southwest. The Steward Observatory reflector would serve as the primary telescope for this proposed facility with specialized equipment furnished by the other two observatories. Such a cooperative effort would allow the relocation of the Steward equipment to a better site and would provide the Perkins and Goethe Link astronomers with access to sophisticated astronomical equipment in a superior location. Although the National Science Foundation was unable to fund this project, the idea of a cooperative observatory led the foundation in late 1954 to appoint a special advisory committee chaired by the noted University of Michigan astronomer Robert R. McMath. This committee would survey the needs of American astronomy and investigate the concept of a national observatory.[21]

During the next two years, preliminary investigations of nearly 150 potential sites in the American Southwest were undertaken by various astronomers under National Science Foundation auspices. The Steward Observatory worked closely with field crews and supplied data from the facility's own site surveys. Carpenter, however, was soon able to provide even more significant assistance to the project. By November of 1955 the survey had determined that Kitt Peak represented one of the best sites for the new observatory. The Tohono O'odham, on whose reservation the peak lay, were hesitant to open this sacred spot for development. Working with Raymond Thompson and Edward H. Spicer of the university's anthropology department, Carpenter invited Tohono O'odham officials to visit the Steward Observatory for a night of viewing and explanation. Shortly after their visit to the Tucson facility in early December, these officials announced that they would recommend to the Tribal Council that a lease be approved for a national observatory test site on Kitt Peak. Tohono O'odham approval allowed the site survey to continue, as did a $545,000 grant from the National Science Foundation.[22]

Although the national observatory site survey was an important interest of the Steward staff, the Tucson astronomers continued to develop their own growing research program. During the late spring of 1956, while engaged in the observatory's continuing supernova research with the Mount Palomar Observatory, Carpenter learned that the National Science Foundation had approved funding to expand these efforts through a cooperative program involving the Steward, Palomar, and Lick observatories. The two-year, $16,000 grant would be administered by the University of Arizona. The foundation earmarked $6,000 for photographic supplies to be shared by the three observatories with the remainder of the grant to provide part-time salaries for observers. Several thousand photographs were taken by the three facilities, adding greatly to astronomers' knowledge of supernovae in distant galaxies.[23]

Staff astronomers continued to pursue their own individual research programs as well. Walter S. Fitch made use of the Steward photometer in his investigation of three variable stars during the mid-1950s. Observing DY Herculis and EH Librae on several nights, he determined light and color curves for both stars. Fitch also noted the close similarity of these stars to VZ Cancri, which he had studied earlier. A more extensive research program focused on the short-period variable CC Andromedae, observed on nineteen nights in 1956 and eight nights the following year. Although the bright night sky of Tucson made accurate observations difficult, Fitch was able to refine many details of the star's periodicities and add to astronomers' understanding of this intriguing variable. His analysis was aided by access to the University of Arizona's new IBM 650 computer. Fitch's investigations of variable stars impressed the editors of the *Astrophysical Journal* and the *Astronomical Journal*, both of whom published his work. For his part, Carpenter pursued spectroscopic investigation of remote double galaxies, identifying filaments of hot gas connecting eighteen pairs of colliding galaxies. His work was complicated by the bright Tucson sky, but led to a published report in the *Publications* of the Astronomical Society of the Pacific. Despite inadequate facilities the Steward Observatory continued to contribute to astronomical knowledge.[24]

Throughout 1956 and 1957, however, astronomical interest in the United States remained focused on the national observatory site survey. Kitt Peak quickly emerged as one of three final locations, while representatives of the National Science Foundation and the recently formed Associated Universities for Research in Astronomy (AURA) signed a $3.1 million contract for the construction, operation, and maintenance of the national observatory. Finally, on 14 March 1958, AURA officials announced that Kitt Peak had been chosen because of its proximity to an academic facility and its superior weather pattern. Although Kitt Peak enjoyed favorable conditions throughout the year, it had especially clear skies during winter and spring, the seasons most likely to produce cloudy conditions on the west coast. Kitt Peak thus complemented the major observatories in California. By the end of August a lease agreement with the Tohono O'odham had been completed and the national observatory site secured.[25]

Although Carpenter and his colleagues expected the national observatory to improve their own situation, the Tucson facility remained, in 1958, a "poor relation" among American observatories. The original members of AURA included such universities as California, Chicago, and Michigan, but Arizona's program was too small to be considered. Carpenter could not quarrel with AURA's decision but he emphasized his disappointment to President Richard A. Harvill in his annual report for 1958. Carpenter wrote that little could be done toward improving the Steward Observatory's reputation while equipment, space, and staff shortages remained crippling. By late December of 1958 this situation changed dramatically. The Kitt Peak project convinced university officials and the Arizona Board of Regents that the Steward Observatory represented an important part of the university community that deserved greater support. To this end, the regents approved proposals to expand campus facilities and staff and to secure a site on Kitt Peak for a Steward Observatory station. Discussions to implement these proposals began at once.[26]

Cooperation between the national observatory and the University of Arizona continued throughout the next few months at a heightened pace.

Carpenter was elected one of three AURA directors-at-large in December and was appointed to three of the organization's committees during the group's Tucson meeting four months later. Colloquia offered by visiting astronomers and frequent specialized conferences were held on the university campus in Tucson. The two observatories also shared library and laboratory facilities. By mid-March 1960 when Kitt Peak National Observatory was dedicated, Carpenter and his colleagues had become an integral part of the growing national observatory. Carpenter, Douglass, and President Harvill were all members of the party that labored up the primitive road to the Kitt Peak site on 15 March for the formal dedication of the new facility.[27]

The astronomy program at the University of Arizona benefited greatly from the selection of the Kitt Peak site. Recognizing the potential importance of astronomy to the growing reputation of the university, officials authorized new faculty and staff as well as a doctoral program. Research seminars were frequently offered by visiting and staff astronomers from the national observatory, providing students and faculty with insight concerning the most recent astronomical advances. Agencies such as the National Science Foundation, the National Aeronautics and Space Administration, and the Office of Naval Research provided needed funds to expand research activity and facilities in a number of different programs. Grants from NASA and NSF proved especially important to the newly established Lunar and Planetary Laboratory, which soon built several large telescopes in the Catalina Mountains north of Tucson and an imposing building on the university campus.[28]

External funding also provided the solution to the longstanding difficulties surrounding the Steward Observatory's campus site. Kitt Peak represented an obvious site for relocation of the thirty-six-inch reflector to make maximum use of the telescope's superior optics. Carpenter initiated a concerted effort to facilitate this move a few months after the formal dedication of the national observatory by submitting a grant proposal to the National Science Foundation to assist in the relocation. Recognizing the value of another large telescope in a superior location, the foundation provided $120,000 to the Steward Observatory. When added to the $100,000 previously authorized by the State of Arizona, this sum allowed Carpenter to begin detailed planning. During 1961, design and construction contracts were negotiated, with a lease for a two-acre site on Kitt

Peak secured from AURA by the end of the year. Construction of the new Steward facility began the following spring and progressed rapidly. The thirty-six-inch telescope was moved to its new home on Kitt Peak in early April 1963. By the end of the summer the Kitt Peak station of the Steward Observatory was operational, fulfilling Tucson astronomers' goal of a major telescope at a superior site.[29]

Ironically, neither of the astronomers who had been instrumental in the early years of astronomy in southern Arizona lived to see the realization of this goal. In March 1962, as the construction of the Steward site on Kitt Peak began, A. E. Douglass died in Tucson at the age of ninety-four. Less than a year later, Edwin F. Carpenter suffered a fatal heart attack in the Steward building on campus. He had returned a few weeks earlier from the Philadelphia meeting of the American Association for the Advancement of Science, at which he was elected a vice president of the association and chairman of the group's astronomy division. Carpenter's status in this important scientific organization was dramatic evidence of the scientific community's recognition of the role to be played by southern Arizona in the advance of astronomy.[30]

Chapter 8 Darwin in the Desert

Arizona's Evolution Controversies

During mid-July of 1925, Arizonans scanned their local newspapers for the latest reports from Dayton, Tennessee. In that small community a young science teacher named John Thomas Scopes was on trial for teaching evolution, violating the recently passed Butler Act that prohibited such activity. The presence at the trial of William Jennings Bryan and Clarence Darrow guaranteed national publicity and kept the Dayton proceedings on the front pages of newspapers for nearly two weeks. The Scopes Trial has remained the most famous example of the antievolution crusade of the 1920s, but it was, in fact, only one aspect of a much more complex phenomenon. Indeed, the Scopes Trial serves best as a symbol of the early years of the controversy concerning the teaching of evolution in America's public schools, a controversy that has continued into the opening years of the twenty-first century. Although Arizona's involvement with this continuing controversy has never attracted the attention of similar developments in the South and in California, the state nonetheless displayed many of the characteristics associated with the national debate.

The antievolution crusade of the 1920s was merely the most dramatic component of a broader movement known as Fundamentalism. Conservative Protestants, uneasy with perceived modernist tendencies in religion and society in the early twentieth century, sought a revitalization of traditional religious attitudes. Although the movement's commitment to a strict biblical literalism has often been overstated, the concepts discussed in the pamphlet series known as *The Fundamentals* (published between 1910 and 1915) clearly sought to re-establish a faith uncorrupted by the higher criticism of Scripture and other modernist ideas. Fundamentalists identified biological evolution with modernism, but the pamphlet writers generally neglected the topic in their essays. Instead, they focused on the need for missionary activity, the danger of textual analysis of the Bible, and the importance of maintaining traditional religious and moral attitudes.

Following World War I, however, the fundamentalist movement became increasingly identified with opposition to the teaching of evolution in American public schools. Noted political reformer William Jennings Bryan, shocked at the carnage in Europe, determined that the underlying cause of the war was Germany's acceptance of a Darwinian "struggle for existence" as a guiding philosophy. Teaching evolution in the public schools of America, Bryan concluded, would point the nation toward the same unacceptable attitudes and away from the reform enthusiasm which had characterized the Populist and Progressive eras. Bryan thus began a campaign to encourage states to ban the teaching of evolution in order to safeguard the nation. Fundamentalists recognized the value of a major figure such as Bryan to their cause, but also acknowledged that evolution represented a challenge to their faith. Securing antievolution legislation quickly became a major activity for fundamentalists and attracted public attention in ways that debates over doctrinal details could not. The passage of Tennessee's Butler Act followed several years of fundamentalist activity in states as diverse as North Carolina, Texas, and Minnesota.[1]

By early 1924 the fundamentalist/modernist clash had clearly emerged in Arizona, as witnessed by the activities of local churches. In Tucson, for example, the Reverend E. C. Tuthill of Grace Episcopal Church was offering a series of sermons on the ongoing debates between fundamentalists and modernists. He attempted to disarm the argument that evolution was the central issue in the controversy by observing that modernists' emphasis on the defense of evolution diverted attention from more significant topics. The key issue, Tuthill concluded, was whether to modernize belief toward the idea that the spirit of truth would lead to truth in all matters of religion, science, and economics. His colleague at the First Christian Church, the Reverend Dr. A. S. Baille, also spoke on the fundamentalist/modernist clash, while the Reverend Richard S. Beal at First Baptist Church offered classes on fundamentalist teachings. Tucson's Congregational Church presented a series of lectures on evolution by local scholars, many of whom were associated with the University of Arizona. These and other activities were reported in the local news media and attracted significant interest.[2]

The comments that attracted the greatest attention in early 1924 were those offered at the 8 January meeting of the Grace Episcopal Women's Service League by University of Arizona astronomer A. E. Douglass. He told the women of the church that science and religion were not incompatible and that the two need not be at each other's throats. At the same time, though, certain biblical accounts could not be taken literally. Joshua's command for the sun to stand still, for example, had to be dismissed; stopping the earth's rotation would have caused immediate destruction. "We must not ask our young people to put faith in things that are obviously to their clear and educated minds untrue," he warned. "We must instead bring our religion down to the present day, not accepting the mythical interpretations given, but giving them instead a common sense religion which will help them and upon which they can lean." Douglass called for a union of science and religion, concluding that when such a union was effected, "I should call it the coming of the millenium."[3]

The involvement of university faculty in discussions of evolution troubled Reverend Beal. Born in Denver, Colorado, in 1887, Beal had studied civil engineering for three years at Colorado Agricultural College before deciding to enter the ministry. Ordained in May 1910, Beal served as pastor of several Baptist churches in Missouri before accepting a call to the First Baptist Church in Victor, Colorado. He served this church between 1914 and early 1918, when he became pastor of First Baptist Church in Tucson. The young preacher immediately began efforts to increase the size of the 200-member congregation, leading the church through several decades of steady growth.

On 10 January 1924, Beal wrote to President Cloyd H. Marvin, objecting to the teaching of evolution on campus. He professed his fondness for the University of Arizona, but emphasized that he was "grieved beyond expression with the bold materialism and gross infidelity of many of its professors." These faculty members should consider alternative explanations to "save from ship wreck on the rocks of infidelity the faith of many of the students." Ten days later, Beal's Sunday morning sermon offered the biblical literalists' perspective in "The Voice of God on the Millenium, an answer to Dean Douglass." The Tucson discussion over evolution, fundamentalism, and modernism continued during the next few months, but by late May the controversy no longer attracted significant attention.[4]

Although the legislature briefly considered a bill to prohibit the teaching of "atheism" and "Darwinism" at the University of Arizona, the state avoided direct involvement with the growing antievolution campaign during the next year. Discussion of Tennessee's Butler Act during the spring of 1925, however, resurrected public interest. Wire service reports began to appear in the state's major newspapers in mid-May, detailing preparations for the Scopes Trial and examining the activities of such prominent antievolutionists as William Jennings Bryan. The specific impact of these events on Arizonans remained ambiguous, as indicated by the meeting of the Arizona Baptist Convention in Mesa. Covered in detail by the *Arizona Republican*, the convention discussed many topics but ignored the evolution issue. *Republican* editor J. W. Spear, on the other hand, offered two editorials on the upcoming Scopes Trial. Observing that Tennessee's Butler Act had been regarded as a joke throughout the United States, Spear stressed that the situation was far more serious now that a trial was looming. The key issues in the case were freedom of speech and freedom of religion, both of which would have to be addressed by local and federal courts.[5]

Readers of the Tucson *Arizona Daily Star* also learned of the upcoming Tennessee case but received a slightly different perspective from editor R. E. Ellinwood. The Tucson editor was intrigued with the announcement that noted attorneys Dudley Field Malone and Clarence Darrow had agreed to defend Scopes. In an editorial entitled "Monkeys, Religion or Both?" Ellinwood argued that personal dislike of Bryan had largely determined the attorneys' decisions. Conducting the trial on "personal grounds" would be unfortunate, he opined, because important issues were involved in the case. From Ellinwood's perspective, the central issue involved the determination of appropriate classroom topics. He argued that no objection would be likely if a teacher were dismissed for teaching anarchy. "That makes it plain," Ellinwood concluded, "that the state has some rights and that the teacher cannot do entirely as he pleases. But where is the line to be drawn?"[6]

Throughout the next two months stories concerning the Scopes Trial appeared on the front pages of Arizona newspapers. Arizonans following the case could also survey local response. In late June *Arizona Republican* editor Spear noted that the potential appearance of scientists as witnesses at the Scopes Trial was largely irrelevant. Science and the accuracy of evolutionary theory were not the issues of note, he argued. The courts were to determine whether the Tennessee legislature had exceeded its authority by proscribing the teaching of a scientific theory. Laws and the Constitution should be the focus of the trial, not a biology textbook. In Tucson, *Star* editor Ellinwood made similar comments. Although religion and science were likely to be the focus of the Scopes Trial, these issues were largely irrelevant. The trial's major issue was a constitutional one: "Has a state legislature the right to decide what shall be taught in the public schools[?]"[7]

As the opening date of the Scopes Trial neared, readers of Arizona's major newspapers learned of the issues and personalities involved and were exposed to analyses by various local observers. A few days before the trial Phoenix resident George Christensen submitted a letter to the *Republican*, responding to press reports that many antievolutionists argued that evolution should not be taught because it had not been "proved." Christensen pointed out that most scientific theories could not be "proved" in the fashion antievolutionists suggested, a conclusion applauded by editor Spear in the same issue. If teachers only taught those concepts that were definitely "proved," the editor argued, "our education would soon be finished." The day before the trial began, R. E. Ellinwood informed his Tucson readers that evolution, science, religion, and other topics were all involved in the Scopes Trial, but the most important aspect of the case remained the fate of democracy. The nation was watching the trial not to find out if Scopes was guilty or to learn the accuracy of evolution. Rather, the trial would show "whether the rights . . . of the individual are to be subordinated entirely to the opinion of the mass." If the Scopes Trial determined that such rights could be deprived "then the rule of the majority will tend to become the tyranny of the majority and to be replaced in time as all tyrannies have been replaced."[8]

On 10 July 1925 *Republican* readers were drawn to the banner headline "Evolution Trial Opens In Tennessee Today." Until the Scopes Trial ended on 21 July, Arizonans read wire service reports on virtually every aspect of the case. Details of the court proceedings, comments from lawyers, and biographical sketches of the participants appeared in multiple stories, usually on the front page of each day's issue. As the trial continued, editorial comments appeared frequently. Stressing that the law itself was on trial, Spear also observed that Bryan's excessively literalistic view of the Bible created a false dichotomy in the case. Evolution was accepted by many scientists and theologians as the working out of God's plan on Earth; Bryan's refusal to consider this possibility clouded the issue. His accusation late in the trial that scientists were part of a conspiracy against the Bible, Spear editorialized, also did little to encourage a peaceful settlement to the controversy.[9]

Coverage of the Scopes Trial in the Tucson press was less extensive but provided Tucsonans with details of the case. Wire service reports consistently appeared on the front page under dramatic headlines, while *Star* editor Ellinwood offered various perspectives during the trial. On 15 July, for example, he suggested that the sensationalist press coverage of the trial would at least educate the public on evolution. From religious leaders, observers would learn that no inherent conflict existed between religion and science. From scientists, the public would learn that evolutionary theory was far more complex than the erroneous idea that humans descended from monkeys. Ellinwood also discussed Europeans' response to the Scopes Trial, stressing their incredulity about the case and the law that precipitated it. Undoubtedly, he suggested, the Butler Act would be overturned, but he closed his editorial by noting, "In the meantime, unfortunately, Europe is judging America by Tennessee." The conviction of Scopes on 21 July ended the trial and removed the story from the front pages. Bryan's death a few days later precipitated significant coverage of the Great Commoner's career, including his participation in the Scopes Trial, but by early August Arizonans were no longer reading about events in Tennessee.[10]

The impact of the Scopes Trial has until recently been misinterpreted. Focusing on the scathing comments of H. L. Mencken, observers concluded that the Dayton trial had, despite Scopes's conviction, been a defeat for the fundamentalist cause. In fact the Scopes Trial encouraged the antievolutionists to expand their efforts. In Tennessee, at least, a legislature had passed an antievolution statute that had been upheld in court. Antievolutionists thus had good reason to believe that their cause could find success elsewhere and campaigned for similar legislation in several states. Mississippi and Arkansas both passed antievolution laws, the latter through the initiative and referendum process.[11]

Arizona's antievolution crusade emerged in the fall of 1927 and focused on the Reverend Richard S. Beal of Tucson. Not yet forty years old, Beal had revitalized First Baptist Church and had recently presided over the dedication of its new building. The church's 1,200-seat auditorium was the largest in the city. In an attempt to lay the foundation for his campaign to secure a statute similar to the Butler Act, Beal invited the noted speaker Dr. Arthur I. Brown to present a series of antievolution lectures. In contrast to most such speakers, Brown's appeal was based on his scientific credentials. A successful surgeon in Vancouver, British Columbia, Brown had been educated at Trinity Medical College in Toronto (M.D., 1897) and had pursued postgraduate work in Scotland. Handbills announcing his talks often described him as "one of the best informed scientists on the American continent," citing his degrees and European training. During the early 1920s Brown became increasingly involved with the antievolution crusade, writing pamphlets and giving speeches that focused on supposed scientific weaknesses in the theory of evolution. He had become such a popular speaker by late 1925 that he took a one-year leave from his medical practice to lecture on science and the Bible. He soon abandoned his medical career completely and devoted the rest of his life to evangelical activity. In addition to traveling widely to give antievolution lectures, Brown also served as field secretary for the World's Christian Fundamentals Association.[12]

By the time of his arrival in Tucson in October 1927, Brown had established himself as the leading scientific "expert" among the antievolutionists, providing the crusade with a weapon to use against the scientific community. Describing evolution as "contrary to the clear facts of Science as well as to the plain statements of Scripture," Brown characterized the theory as "the greatest hoax ever foisted on a credulous world." In addition to giving his lectures at Tucson's First Baptist Church, Brown provided ample material for the local media. In an interview with the *Arizona Daily Star* for 11 October, Brown stressed that Genesis remained a valuable scientific work. This book gave a definite order to creation, an order that modern scientists could not refute. Genesis also provided the basis for rejecting the concept of the evolution of species "because God commanded that everything living must reproduce itself 'after his kind.'" Brown continued, "This has been denied by those who would prefer the mythical evolutionary process, but thousands of experiments have shown that this Supreme Law cannot be broken." A long letter to the *Star* two days later expanded his comments on the "decrepit and tottering theory" of evolution.[13]

Brown's lectures and media activity encouraged Beal in his campaign to secure an antievolution statute for Arizona. Midway through Brown's lecture series, Beal announced that a survey of Arizona ministers revealed that more than half were sympathetic to his campaign. Although the details of his survey remained unclear, the Tucson minister stated that he would launch a statewide campaign to secure appropriate legislation and would bring in various speakers to aid in his efforts. Brown's lectures and Beal's comments continued to insist that evolution was not scientific and in fact represented mere guesswork. *Star* editor R. E. Ellinwood addressed this aspect of the controversy in his editorial of 17 October, emphasizing that referring to evolution as a "theory" did not mean that scientists rejected it. Scientists also referred to "gravitational theory," he noted. Further, the correctness of the evolutionary explanation was not necessarily the most significant consideration. More important was the scientist's "freedom to search unhampered for the truth, to present the facts learned to others, and leave them free to make their own decisions." The effort to outlaw the teaching of evolution threatened that freedom and represented "a monstrous contradiction of the principles upon which our nation was founded."[14]

The *Star* conducted its own poll of Tucson clergy concerning Beal's proposed legislation, announcing on 19 October that seven of the eleven ministers who responded opposed the measure. These opponents argued the necessity of church/state separation and also stressed that no necessary contradiction existed between science and religion. Beal dismissed the results of this poll, repeating his opposition to teaching evolution in tax-supported schools because it was a system "which tends to destroy the soul and wreck the faith of our boys and girls." Indeed, the supporters of evolution were engaged in a campaign against the church. "There can be no question," Beal asserted, "but that this theory of beast ancestry is one of the most effective weapons ever invented by the devil to wreck the essentials of Christianity." Despite Beal's efforts, Tucson showed little enthusiasm for the proposed antievolution law. The Presbyterian Synod meeting in Tucson overwhelmingly passed a resolution opposing the proposed statute, while various sermons and letters to newspaper editors criticized the effort. By mid-November, Beal had abandoned his campaign; Arizona would not witness a legislative fight over the teaching of evolution in the public schools.[15]

Brown's Arizona visit was not restricted to the Tucson lecture series arranged by Reverend Beal. Beginning on Monday, 31 October, he presented his "Science and the Bible" lectures at Capitol Methodist Church in Phoenix. Brown's nightly lectures continued throughout the week, although his activity on Sunday, 6 November, included two additional lectures during the day and a morning address to the Radio Community Bible Class broadcast over station KFCB. Offering discussions very similar to those presented in Tucson, Brown provided lectures on such topics as "Genesis and Modern Science," "Wonder Book of the Ages—the Bible," and "Men, Monkeys and Missing Links—The Truth About Evolution." Addressing the last topic in his Friday night lecture, Brown characterized evolution as unreasonable, unscientific, and unscriptural. He observed that individuals who truly understood science were rejecting evolutionary concepts because of the many weaknesses in the theory, the most serious of which was lack of observational proof. "Evolution depends on and demands numberless changes of species," he argued, "but there has never been a single instance of this absolutely necessary phenomenon." Brown followed his presentations at Capitol Methodist with several lectures during the second week of November at McCahan Desert Chapel north of Phoenix, whose congregation included many of the health-seekers living nearby. Although Brown's Phoenix visit failed to arouse significant interest in an antievolution crusade, his appearances in Arizona identified the state with the continuing national campaign to prohibit the teaching of evolution in the public schools.[16]

The end of the national antievolution crusade in the late 1920s removed the topic from public view but it did not return evolution to the public school curriculum. In many states local school boards directed teachers to ignore the topic to avoid controversy. More damaging to science education, however, was the decision by textbook publishers to eliminate or trivialize the discussion of evolution in order to maintain sales. As a result, students in the United States for more than three decades after the Scopes Trial learned little about the fundamental concept of modern biology. The poor state of science education proved a widespread problem, as other disciplines were no better taught than biology. The Soviet Union's successful launch of the first artificial Earth satellite in October 1957 served as a dramatic revelation of the implications of this situation. Before the American satellite program achieved success with *Explorer I* the following January, the Soviets had launched a second satellite (with its canine passenger) and had forcefully challenged American preeminence in science and technology. The lack of effective science education in the United States was an obvious factor in the nation's loss of scientific status.

Although the Eisenhower administration downplayed the seriousness of the "Sputnik crisis," science educators stressed the need for curricular improvements. With funds from the National Science Foundation, teams of scientists and educators began writing improved textbooks and designing educational material to correct the dismal state of science education. The most dramatic changes were in the field of biology, through the endeavors of the Biological Sciences Curriculum Study (BSCS). The biologists and science educators who participated in this activity produced three separate texts, each designed for a general tenth-grade biology course. The most traditional text (known as the Yellow version) focused on cells, development, and evolution, while the Blue and Green versions emphasized, respectively, biochemistry and ecological perspectives. Emphasizing student activities and the most recent scientific concepts, the BSCS curriculum sought to present science as a way of knowing about the world, rather than as a list of "facts" to memorize and regurgitate on examinations.[17]

Arizona quickly emerged as an important focus of the new science education. During the 1960–61 academic year, Phoenix served as one of the test centers for the BSCS Blue version. Although this textbook did not focus on evolution as did the Yellow version, evolutionary concepts played an important role in the material. A few parents and ministers objected to this evolutionary emphasis, but a significant controversy did not erupt until the following academic year. Readers of the *Arizona Republic* for Sunday, 28 January 1962, learned of a letter from three prominent lay leaders of the Mormon church to the Phoenix superintendent of schools. Despite the lack of an official Mormon position on evolution, these leaders objected to the teaching of this topic in the public schools. Criticizing evolution as guesswork, they argued that teaching evolution "is a very dangerous situation and in view of the fact that the theory has not been established as fact, we think it should not be taught in the schools." The Mormon leaders appeared more concerned, however, with religious questions. Stressing that evolution was "in direct opposition" to Christian teachings, they told the school superintendent that including the concept in biology education "comes as close to teaching atheism as one can at the secondary school level." The BSCS Blue version, they concluded, should be removed from the curriculum because it taught evolution.

The response of school superintendent Howard C. Seymour provided BSCS supporters with little encouragement, although he rejected the suggestion that the Blue version be discarded. He stated that the BSCS material was being used in an accelerated program offered as an experiment and that no student was required to take the course. Seymour also played down the impact of evolution in the curriculum. Referring to the instruction in evolution, he told an *Arizona Republic* reporter: "Students are not expected to believe this. The instruction is only one of several attempts to explain the origin of life," he continued, "and no attempt is being made to supercede any family or religious instruction." The Phoenix controversy over the BSCS Blue version soon subsided.[18]

As BSCS material became more visible in the curriculum in 1963, however, its evolutionary perspective precipitated a more heated response. Phoenix radio station KRUX aired a series of strong editorials criticizing BSCS during the spring of that year, but it was the November meeting of the state board of education that led to Arizona's most dramatic confrontation with evolution during the 1960s. Several Arizonans came before the board to protest the teaching of evolution in the state's public schools. Phoenix parent Harold Bates, whose children had local district permission to leave the classroom when evolution was discussed, asked if the board possessed the power to compel teachers to grant such absences. The board attorney suggested that this was a matter for individual districts to decide. Perceiving such a response as an avoidance of the issue, Pastor Aubrey L. Moore of the West Van Buren Baptist Church in Phoenix openly criticized the board for refusing to take a stand against including evolution in textbooks. "If you won't do it," he warned the board through an *Arizona Republic* reporter, "we're going to get people together who will do something about it."[19]

Moore, who had moved to Phoenix from Mississippi in 1955, had recently orchestrated the repeal of the Phoenix housing code through a successful petition drive and special election. Despite the opposition of both major Phoenix newspapers, voters overwhelmingly repealed the code and made it impossible for the city to implement an urban renewal program. Moore's comment to the board seemed to imply that a similar petition drive might be organized to force evolution out of the curriculum. The editor of the *Arizona Daily Star* in Tucson certainly interpreted Moore's comments in this fashion, criticizing the pastor's actions as "a publicity stunt" and warning readers that Arizona could look extremely "foolish" if the state hosted another Scopes Trial.

Despite the possibility that Arizona might appear foolish, Moore continued his attacks on evolution. Shortly before the 18 December board of education meeting, he told reporters that teaching evolution was the first step toward communism. He also orchestrated a request to the board to remove textbooks discussing evolution from the public schools. During the meeting, at which Moore was ordered to dismantle tape recording equipment he had brought with him, the board refused to ban such texts. The attorney general had recently issued an opinion that the board possessed only limited power to prevent the teaching of evolution and that any decision to use this power should follow a court case. In addition, the attorney general advised that the board could not alter the list of textbooks approved in 1962 for a five-year period.[20]

As expected, Moore immediately announced plans for a petition drive to place an antievolution proposition on the ballot for November 1964. Proposed as a constitutional amendment, Moore's initiative was defined as an act to prevent the teaching of atheism in the public schools. The proposed amendment, however, defined atheism as the "teaching of any theory that denies the existence of God and the Divine creation of man in God's image," and as the teaching "that man evolved from a lower order of animals." As the petition drive neared its July deadline, Moore became increasingly active. He emphasized in interviews and speeches that the Butler Act was still in force in Tennessee, showing that antievolution statutes remained viable. He visited a social science class at Camelback High School and was greeted with amusement and incredulity by students. Moore cited his reception as proof that Arizona children were being "brainwashed" in favor of Darwin and against the Bible.[21]

Opponents of Moore's campaign were also active. The American Civil Liberties Union stressed that the supposed focus on preventing the teaching of atheism masked the antievolution purpose of the proposed amendment. The Arizona Academy of Sciences issued statements opposing the measure and contacted religious leaders throughout the state to encourage them to speak out against Moore's initiative. The Phoenix Presbytery voted to oppose the proposed amendment and urged congregations not to sign Moore's petitions. The president of the Arizona Ministers Council of American Baptist Churches wrote a letter to the editor of the *Arizona Republic*, recording his opposition to this latest antievolution effort. Forty-three Methodist ministers and the president of the Phoenix Rabbinical Council issued a public statement urging Arizonans not to sign Moore's petition.[22]

The Phoenix pastor was unmoved by his colleagues' opposition. He responded to this last statement by telling a reporter, "It was the Jews who crucified Christ. Jews don't believe in the Bible. And neither do those hypocritical Methodist ministers." Those individuals who accepted the idea that evolution might well be part of God's plan were similarly dismissed. "The people who call themselves theistic evolutionists," he informed a *New York Times* reporter, "don't know what they're talking about. There's nothing in the Bible about a fish turning into a man." Despite his efforts, Moore's petition failed to secure enough signatures to be included on the November ballot. His plans to take the issue into the courts and to begin another initiative effort for the 1966 election failed to materialize.[23]

Shortly after the failure of his initiative campaign, Moore told a Tucson reporter that his purpose was not actually to ban evolution but only to secure the teaching of divine creation at the same time. Given Moore's actions and statements over the past several months, his comment appears curious if not suspect. Yet his call for "equal time" for creation represented a tentative new path for the antievolution crusade that had already led to an intriguing Arizona development. On 20 February 1964, as Moore was organizing his petition drive, Representative James F. E. Young of Maricopa County introduced House Bill 301. His bill would require "equal time" for the "Doctrine of Divine Creation" in the public schools if evolution were taught. The bill was referred to the Committee on Judiciary, Education, County Affairs, and Boards and Commissions, who returned a majority report in favor of consideration by the full house. Curiously, the committee retained House Bill 301, a procedure that would require the house to retrieve the bill specifically for consideration. The house declined to take such action. A similar bill introduced in the senate during February 1965 died in the education committee. Although neither Arizona bill made a significant impact, the concept of "equal time" would prove to be an increasingly important focus of the continuing campaigns to remove or compromise the teaching of evolution in the public schools.[24]

In retrospect, the 1960s marked the end of the antievolution crusade that had surfaced a half-century earlier. In addition to the inclusion of evolution in the biology curriculum, the decade witnessed the removal of antievolution statutes. Faced with a potential court case involving a teacher who had been dismissed for discussing evolution, the Tennessee legislature repealed the Butler Act in 1967. The U.S. Supreme Court invalidated the Arkansas statute the following year in the famous *Epperson v. Arkansas* decision, while the Mississippi Supreme Court ruled that state's antievolution law void in 1970. In an age characterized by science and technology such laws appeared anachronistic.

Legislative and judicial action, however, failed to eliminate opposition to the teaching of evolution. Indeed, the same decade that witnessed the end of the antievolution crusade marked the opening of the next campaign against such teaching. Led by evangelical Christians trained in science and engineering, a new movement emerged in the late 1960s that claimed to offer a scientific alternative to evolutionary explanations of life's origin and development. Variously denominated "creation-science" or "scientific creationism," this new perspective offered an explanation of life on Earth that left intact literal biblical interpretations. Although the creationists' publications stressed the scientific legitimacy of their ideas, most of their efforts were aimed toward discrediting evolutionary orthodoxy. The supposed lack of transitional forms in the fossil record, the failure of scientists to show one species changing into another, and questions concerning radiometric dating techniques were merely a few of the arguments creationists employed to discredit evolution. If, therefore, both theories (creationism and evolution) were based on untestable assumptions, neither could claim favored status. Both theories were equally viable. Both theories should be taught in the public schools.

Throughout the 1970s opponents of evolutionary teaching campaigned for "balanced treatment" of creation and evolution. The most straightforward method to achieve this goal was to provide for "equal time" for creation-science in the curriculum. At the local and state level, education boards were deluged with publications from creationist organizations which offered explanations of life's origin and development that appeared sound to individuals with limited science background. Legislators received similar material and were lobbied by individuals and groups who sought state laws to require local schools to offer the creationist alternative. This effort was particularly successful in Arkansas, where the legislature passed an "equal time" bill in the spring of 1981. Stressing the supposed scientific nature of creation-science and raising such issues as "fairness" and "freedom of choice," proponents of the legislation had little difficulty gaining passage of this measure. The legal challenge emerged at once, however, as opponents of the legislation emphasized that creation-science was little more than thinly disguised Protestant fundamentalism and as such was an effort to establish a specific form of religion. Such establishment, as had been shown in numerous U.S. Supreme Court cases, was patently unconstitutional. This conclusion was affirmed by U.S. District Judge William R. Overton who ruled in the *McLean v. Arkansas* decision of January 1982 that the primary purpose of the Arkansas legislation was the advancement of religion. Although calls for "equal time" and efforts to enact viable legislation toward this end continued, the Establishment Clause of the First Amendment appeared to present an insurmountable barrier to the creationist campaign.[25]

In Arizona, nonetheless, the "equal time" concept continued to attract attention. In early February 1982 Rep. Jim Cooper, a Republican from Mesa who chaired the House Education Committee, introduced a bill to achieve balanced treatment. Evidently accepting the concept that evolution and religion were incompatible, Cooper drafted his bill to require that when evolution was taught in a public elementary or high school the teacher "shall not present the theory of evolution in such a way as to foster a belief in a religion or [to] cause a disbelief in religion." The state board of education was directed to establish rules and regulations for local school boards to use in establishing evolution courses. Teachers violating this law could be fined up to $10,000 and could be imprisoned for up to one year. The bill was initially assigned to the House Committee on Environmental Affairs, but the committee chair's opposition to the bill played an important role in preventing the measure from reaching the house floor. The recent Arkansas case, which seemed to show that such laws could not survive constitutional challenges, also convinced legislators that the Cooper bill should be avoided. As was the case in several states in 1982, Arizona's "equal time" bill failed to secure serious consideration at the legislative level.[26]

The issue itself, however, attracted significant attention. As the legislature was considering Cooper's bill, the Baptist Student Union at the University of Arizona sponsored a debate between two evolutionists and two of the leading figures in the creation-science campaign, Henry Morris and Duane Gish. Morris, an engineer, and Gish, a biochemist, had published numerous books and articles articulating the creation-science perspective and had traveled throughout the United States for lectures and debates. The evolutionist perspective would be presented by entomologist David Milne of Evergreen State College (Olympia, Washington) and biologist Kenneth Miller of Brown University. The 12 February debate attracted some 300 people to the Tucson campus, but the performance offered little beyond standard arguments. Miller attempted to counter the charge that evolution directly challenged the existence of God by observing that "Science doesn't deal with divinity." Morris provided a concise summary of the goals of the creation-science movement when he announced, "We're concerned with winning a hearing. For a long time, evolution has dominated in schools. It's time to reconsider creationism as an alternative to evolution in schools and universities." Milne discussed the number and significance of transitional fossils, while stressing that evolution was a "morally neutral process."

As he frequently did on such occasions, Duane Gish attempted to give a witty presentation, at one point dismissing evolution as the "fish to Gish" theory. He also questioned the Big Bang explanation of the origin of the universe, telling the audience, "I find the proposition that you and I, with 12 billion brain cells, have been produced from a cloud of hydrogen preposterous." Morris rested the creationist case by noting that both creationism and evolution were based on faith, the latter having become "a scientific religion." Repeating arguments that would have been familiar to William

Jennings Bryan or Arthur I. Brown, Morris announced that the religion of evolution was losing converts. Many scientists were now turning to creationism, convinced that insufficient evidence existed to support evolutionary explanations. Morris further attempted to argue that his evolutionist opponents were attacking the biblical creation model rather than the scientific creation model, a distinction that the audience found both puzzling and suspect. Although the debate failed to advance the creationist cause in Arizona in early 1982, such actions kept the controversy before the public.[27]

The teaching of evolution in the public schools continued to concern the state legislature as well. In April of the following year, the Arizona Senate voted 17–13 in favor of a bill to require that ideas concerning the "origin of man" be taught as theory and not fact. Recognizing that *theory* was frequently interpreted as *guess* by nonscientists, creationists had increasingly resorted to this tactic to weaken the status of evolution in the public school science curriculum. The senate's decision to pass this bill was primarily a result of the actions of House Education Committee chair Jim Cooper, who held hostage a priority senate bill that would require school districts to adopt specific retention and promotion guidelines. Once the senate passed the measure, the house quickly followed suit. On 29 April, Governor Bruce Babbitt vetoed the legislation, arguing that it violated the First Amendment and was an apparent attempt to limit the scientific study of evolution. The legislature failed to override the veto.[28] Arizona creationists would have to develop alternative tactics to achieve their goals.

Occasional efforts to draft workable "equal time" bills continued, as shown by another incident involving Jim Cooper in early 1987. Now serving as education advisor to controversial Governor Evan Mecham, Cooper appeared before the House Education Committee on 4 February to testify in favor of House Bill 2095, another "equal time" measure. During his testimony a committee member asked Cooper what would happen if a student told a geography teacher that his parents had told him that the world was flat. Did not the teacher have a responsibility to provide a correct explanation, whether in geography or biology? Cooper's response surprised many. "The schools," he told the committee, "don't have any business telling people what to believe." The bill died in committee. This action appeared particularly wise a few months later when the U.S. Supreme Court announced its decision in *Edwards v. Aguillard*. The Court found that the Louisiana "equal time" statute passed several years earlier was a violation of the Establishment Clause of the Constitution.[29]

Antievolutionists of the late twentieth century enjoyed significant public and political support throughout the nation despite their lack of success in the legislative and judicial arenas. Opinion polls indicated that the public remained comfortable with the concept of teaching both evolution and creationism in the public schools. Such polls caught the attention of school boards and legislatures, who often pursued appropriate curricular changes. Although New Mexico scuttled its three-year-old science standards in 1999, thus restoring evolutionary concepts to the science curriculum, other states such as Louisiana, Oklahoma, Nebraska, and Kansas witnessed efforts to minimize the place of evolution in the science classroom.[30]

The continued support for the antievolution perspective was at least in part an artifact of a broader cultural change during the closing years of the twentieth century. The growing rejection of expertise led clearly to an antiscience mindset that worked to the advantage of antievolutionists. From this perspective, science deserved no special status, as it was little more than one of many "belief systems" available for consideration. Scientific explanations and results could thus be rejected if they clashed with other "beliefs." The contentious debates concerning genetically modified crops, for example, rarely focused on the laboratory and field tests that indicated little potential risk from these plants. Similarly, Native American activist Vine Deloria Jr. attracted much publicity in 1995 when he published *Red Earth, White Lies: Native Americans and the Myth of Scientific Fact*. Arguing that anthropology was an illegitimate field of endeavor, he stressed that white scientists could not possibly understand indigenous cultures. Much more reliable than anthropologists' explanations were the myths and traditions that had guided Native Americans for centuries. That such myths were by definition outside the scientific mainstream did not strike Deloria as important. After all, he wrote, the recent collision between the comet Shoemaker-Levy and the planet Jupiter clearly indicated that "the era of uniformitarian orthodoxy must come to an end." Such views played a major role in the protracted controversy concerning skeletal remains discovered in Washington state in 1996. Although a careful study of these ancient remains would undoubtedly provide valuable insight concerning the early settlement of the North American continent, federal agencies and courts severely restricted such study in favor of local tribal claims for reburial under the 1990 Native American Graves Protection and Repatriation Act. The suggestion that analysis of the remains would add to knowledge failed to impress those who rejected the anthropologists' perspective. "We already know our history," noted Armand Minthorn of the Confederated Tribes of the Umatilla Indian Reservation, the likely claimants to the remains. "We were created here. We didn't cross any land bridge. That's what our history tells us."[31]

In the Southwest, similar rejection of scientific analysis and conclusions characterized the late twentieth century. Biologists' studies of the red squirrel habitat on Mount Graham, Arizona, suggested that the proposed observatory on that site would have minimal impact on the species. Such conclusions were rejected by some environmentalists who opposed the project and rarely impressed judges who heard the various suits brought to halt construction. Geological and other studies of the Waste Isolation Pilot Project site near Carlsbad, New Mexico, failed to remove fears that the storage of nuclear waste represented a potential disaster. Originally authorized by Congress in 1979, the project did not receive its first shipment of waste until March of 1999, after the expenditure of more than $2 billion and innumerable court and administrative proceedings. Despite approval by the Environmental Protection Agency of plans developed with advice from such groups as the National Academy of Sciences, opponents continued to pursue court challenges based on procedural grounds for nearly two decades.[32]

Arizona witnessed the rejection of expertise as well, often through the continued effort to compromise the teaching of evolution in public schools. The state committee appointed to draft a "science essential skills" document included nationally known creationist Walter T. Brown Jr. The state's decision to authorize "charter schools" provided the opportunity to teach creation-science despite the legislative requirement that such schools be nonsectarian. The Academic Standards for Science adopted by the state board of education in 1997 failed to include evolution as one of the "unifying concepts and processes" guiding science. This decision appeared particularly instructive concerning the nebulous status of evolution in

Arizona public education, as the state policy reflected the National Science Education Standards in all other matters except the inclusion of this concept.[33]

As the twentieth century closed, therefore, Arizona continued to participate fully in the national debate over the teaching of evolution in the public schools. Although the state never passed an antievolution statute and never achieved the dubious distinction of defendant's status in a major court case, the issues raised in other states were present in Arizona from the 1920s through the 1990s. Scientists, theologians, and politicians were involved in the resulting debates, as were concerned residents of various backgrounds. The dramatic growth of Arizona's scientific reputation during these decades failed to eliminate opposition to the teaching of the most fundamental concept of modern biology. In this, as in much else concerning the evolution controversy, Arizona provides a disconcerting perspective on the role of science in American culture. That Arizona shares this perspective with the rest of the nation makes it no less troubling.

Chapter 9 The Southwest and Interplanetary Exploration

During the last four decades of the twentieth century, American space probes surveyed all but one of the planets of the solar system. Data from these missions significantly altered scientists' views of the structure and evolution of the solar system, providing new knowledge of planetary satellites, atmospheric characteristics, and the dramatic rings of the outer planets. Although overshadowed by the better publicized manned space program, NASA's interplanetary probes returned far greater benefits to the scientific community.[1]

Even before the dawn of the space age, the American Southwest had been a focal point for the study of the solar system. Fascination with the planet Mars led to the foundation of the Lowell Observatory in 1894, after which astronomers at the Flagstaff facility expanded their research to include other planets. Mars was also the focus of early astronomical efforts at the University of Arizona despite the school's lack of a large telescope. The completion of the university's Steward Observatory in the early 1920s greatly expanded the region's astronomical equipment, as did the establishment of the McDonald Observatory in west Texas in the late 1930s. Although the telescopes of these facilities were employed in a wide range of research endeavors, the planets of the solar system remained an important topic for the region's astronomers.

Building on this foundation, scientists in the American Southwest played a leading role in NASA's exploration of the solar system. From the early lunar probes of Project Ranger to the sophisticated hardware of *Mars Pathfinder*, the region's scientists participated in mission planning, instrument development, and the interpretation and dissemination of experimental results. Their presence in the development and progress of these missions paralleled the growing institutional base of science in the Southwest and encouraged the expansion of that base in the years that followed.

Toward the Moon: Project Ranger

The successful launch of *Sputnik I* in October 1957 shocked most Americans. Numerous observers suggested ways to reassert American dominance in science and technology, but "beating" the Russians to the Moon soon emerged as the primary goal. William H. Pickering, director of the Jet Propulsion Laboratory (JPL) at the California Institute of Technology, suggested that existing JPL work on space probes could serve as an important first step toward the Moon. By late 1959 the newly formed National Aeronautics and Space Administration (NASA) had directed the laboratory to undertake a series of lunar probe missions, the details of which were developed during the following year.[2]

Although the Washington-Pasadena axis defined the early years of the unmanned space program, American scientists from other institutions were also involved. Among these noted figures was the Dutch-born astronomer Gerard P. Kuiper, who served on the NASA Working Group on Lunar Explorations. Kuiper had joined the staff of Lick Observatory after receiving his Ph.D. from the University of Leiden in 1933, moving to Harvard in 1935 and to Yerkes Observatory and the University of Chicago in 1936. Following various science-related activities during the war, Kuiper returned to Yerkes where he became increasingly interested in planetary and lunar studies. By the mid-1950s he had developed plans for a comprehensive photographic lunar atlas and had begun to consider the prospect of a lunar and planetary studies institute.[3]

Kuiper's position at Yerkes Observatory, however, presented numerous

Gerard P. Kuiper established the Lunar and Planetary Laboratory at the University of Arizona and played an important role in NASA programs during the 1960s. (Courtesy of the University of Arizona Space Imagery Center)

barriers to his proposed institute. Administrative duties as observatory director were increasingly time-consuming, while his colleagues tended to dismiss planetary studies in favor of stellar and galactic work. By the late fall of 1959 Kuiper had determined that a new location for his studies was necessary. The recent selection of Kitt Peak as the site for the national observatory suggested an appropriate opportunity. Kuiper, observatory director Aden Meinel, and various officials at the University of Arizona quickly worked out the details of Kuiper's move to Tucson to establish a planetary institute. The university would gain another noteworthy science facility and the promise of external funding to support Kuiper's work. Kuiper resigned as Yerkes director early in 1960, just before his "Photographic Lunar Atlas" was published in April.[4]

NASA's continuing support for lunar and planetary studies was an important aspect of the anticipated external funding for Kuiper's Tucson facility. By the time he and his staff began their move to Tucson during the summer of 1960, NASA and planetary astronomers had determined that federal support for ground-based astronomy would be a crucial part of NASA's lunar and planetary work. The Southwest quickly emerged as an important base for this activity. Kuiper secured funding for several telescopes to support his Lunar and Planetary Laboratory, including a sixty-one-inch reflector in the Catalina Mountains north of Tucson. Other facilities in the Southwest, such as the Lowell Observatory in Flagstaff and New Mexico State University in Las Cruces, also obtained generous NASA funding for ground-based astronomy.[5] Geologist Eugene M. Shoemaker, who had spent the 1950s studying various terrestrial craters from volcanoes, meteor impacts, and nuclear test explosions, guided the establishment of the U.S. Geological Survey's astrogeology branch in Flagstaff. Shoemaker's goal was a better understanding of lunar craters, a goal that meshed nicely with the expanding American space program and led to significant NASA funding.[6]

Kuiper, Shoemaker, and their colleagues were increasingly involved in the initial NASA lunar mission called Project Ranger. An extension of earlier plans developed by the Jet Propulsion Laboratory, the *Ranger* probes were originally planned to deliver instruments to the Moon to determine the characteristics of the lunar surface. As planning and design for the first five *Ranger* probes progressed, however, questions concerning the goals, management, and organization of the program constantly arose. A reorganization of the program in January 1961 was followed a few months later by a reorientation of Project Ranger as a result of President John F. Kennedy's call for a manned lunar landing. Now defined as "direct support" activities for Apollo, the Ranger missions would be far less focused on science, much to the chagrin of many at JPL and throughout the program.[7]

Despite these and other problems, planning for Project Ranger advanced during 1961. Kuiper and Shoemaker were active in this planning and by October were appointed to the team of experimenters who would interpret television pictures broadcast back to Earth before the probes made impact with the Moon. Kuiper was soon appointed principal investigator of these experiments. He was also serving in a similar capacity in the Surveyor program to land spacecraft on the Moon and was a member of the Scientific Working Group for the Apollo program. Indeed, Kuiper was involved in so many lunar and planetary projects in late 1961 that NASA headquarters wondered if he and his laboratory might be over-extended. This concern failed to change Kuiper's status within NASA, however, as his expertise remained crucial to the agency.[8]

The potential value of the Ranger program appeared somewhat problematic by the end of 1962. The first five spacecraft, launched between August 1961 and October 1962, suffered various equipment failures and provided little information. Following these failures both NASA and the Jet Propulsion Lab convened review boards. Criticisms were widespread. The laboratory was faulted for inadequate systems design review, weaknesses in design and assembly, a lack of focused project management, and a "shoot and hope" approach that undermined design excellence. NASA officials urged that the program adopt clearer and simpler mission objectives, that the agency itself monitor the program more closely, and that pre-launch testing of the spacecraft be increased. By mid-December new guidelines were in place. The Ranger program now had a single objective: television pictures of the lunar surface to provide information for the Apollo program. All other experiments and instruments were to be sacrificed, with the weight saved to be used for redundant engineering features to insure that the spacecraft would function as designed. The Block III spacecraft (*Rangers 6* through *9*) would reflect this new focus.[9]

Kuiper, Shoemaker, and their colleagues were actively involved in this process. Kuiper remained principal investigator for the television experiments on these probes and, with Shoemaker, played an important role in the *Ranger* modifications. At a meeting of the television experimenters in Pasadena in late November, 1962, for example, both scientists emphasized that extensive testing of the television subsystem was essential before launch. They also urged more active participation of scientists in future testing and evaluation of this system. The importance of television imaging to the Apollo program made such recommendations of great interest to NASA officials whose concentration on the manned lunar landing increasingly defined the American space program.[10]

The need for a successful mission had a significant impact on Project Ranger. Throughout 1963, as many as 900 people were involved in the reengineering of the *Ranger 6* spacecraft. Closer attention to details revealed various problems in both the Atlas launch vehicle and the spacecraft itself, delaying the planned launch date to late January 1964. Ironically, Ranger's redefinition as an Apollo support mission had, by the end of 1963, minimized its importance. *Ranger* was designed to reveal the nature of the lunar surface, especially as it related to the design for the *Apollo* lander. The earlier delays and failures in the Ranger program had forced Apollo engineers to freeze the landing gear design without the anticipated Ranger data. The assumption of a solid lunar surface underlay these decisions, leaving Ranger with little to contribute to the lunar program.[11]

The launch of *Ranger 6* on 30 January 1964 went on as scheduled, as would the remaining three missions. Details of potential landing sites remained uncertain and could have an important impact on the Apollo missions. Unfortunately for all involved, *Ranger 6* was another disappointment. After a perfect flight to the Moon, the television cameras failed to function. Inquiries immediately began at NASA headquarters, the Jet Propulsion Laboratory, and in Congress. Once again, design flaws, management shortcomings, and inadequate testing procedures were charged. To salvage the program, NASA and JPL further modified design and testing procedures in an effort to make *Ranger 7* a reliable spacecraft.[12]

Analysis of earlier failures and modifications to the new spacecraft involved scientists and engineers for the first half of 1964. Because the television subsystem had become the focus of the mission, Kuiper, Shoemaker, and Ewen A. Whitaker (Kuiper's colleague at the Lunar and Planetary Laboratory who had been increasingly active in Ranger site selection decisions) carefully monitored *Ranger 7* development. When the spacecraft started its journey toward the Moon on 28 July, all three scientists were in Pasadena. Three days later, after a nearly flawless mission, *Ranger 7* hit the Moon within twelve kilometers of its target point. This time the cameras functioned as designed, returning more than 4,300 pictures before the spacecraft crashed into the lunar surface.[13]

Kuiper and his colleagues had little time to savor this long-awaited success. Shortly after nine o'clock that night, less than fifteen hours after impact, they addressed a press conference in the JPL auditorium. Their discussion of the mission and the numerous slides they displayed provided reporters with a clear indication of both the scope of the mission and the fact that the lunar surface appeared to be safe for manned landings. Expressions of relief over the successful mission were also evident in Pasadena. Kuiper, as principal investigator, opened the press conference with the comment, "This is a great day for science, and this is a great day for the United States. What has been achieved today is truly remarkable."[14]

Although the immediate analysis of *Ranger* data attracted much atten-

This photograph, taken by *Ranger 7* on 31 July 1964, was the first image of the Moon taken by an American spacecraft. The spacecraft transmitted more than 4,000 images before its impact on the Moon seventeen minutes later. (Courtesy of the University of Arizona Space Imagery Center)

tion and interest, the scientists involved in the mission had more work to do. Exhaustive evaluation of these data consumed scientists' energies for the next month. The *Ranger 7* Preliminary Scientific Results Conference held in Washington on 28 August provided a much more complete survey of the new lunar knowledge. Kuiper and Shoemaker interpreted *Ranger* results as indicating that the lunar "seas" were lava flows and would provide a reasonably solid surface. Harold Urey, the Nobel Prize–winning chemist who was also a member of the Ranger experimenter team, remained unconvinced that the surface was solid. He maintained that the surface could be spongy enough to cause landing difficulties. Other information concerning the Moon was more certain. Although the lunar surface was much more heavily cratered than earth-based observations had indicated, the surface itself was both smoother and less sloping than some astronomers had feared. Later meetings of Ranger scientists at the Lunar and Planetary Laboratory in Tucson (23–24 October) and at NASA headquarters in Washington (22 December) failed to resolve all the differences concerning the interpretation of the *Ranger* photographic results. When the scientists' final reports were submitted to NASA headquarters in January of 1965, however, the assumption of a lunar surface with sufficient strength to support the *Apollo* lander was widely accepted.[15]

The success of *Ranger 7* provided scientists with much information, but the program included two more missions that were likely to add to the store of knowledge. Kuiper, Shoemaker, and Whitaker remained active in the planning for these flights. The mid-February flight of *Ranger 8* was another success, returning more than 7,000 pictures and confirming that the basic surface structure of the Sea of Tranquility was essentially the same as that of *Ranger 7*'s target in the Sea of Clouds. Because the previous two missions had removed most uncertainties concerning the *Apollo* landings, Kuiper and his colleagues were able to persuade mission directors to target an area of greater scientific interest for *Ranger 9*. Ewen Whitaker was instrumental in choosing the interior of the crater Alphonsus in the lunar highlands. *Ranger 9* crashed into the Moon on 24 March within five kilometers of its aim point, returning live television pictures of its journey. Various meetings and discussions about the results of the last two *Rangers* took place throughout the remaining months of 1965, with a final published report appearing in March of the following year.[16]

The involvement of southwestern scientists such as Kuiper, Shoemaker, and Whitaker in the Ranger program had an important impact on the region's scientific reputation, especially as it related to later lunar missions. Eugene Shoemaker served as principal investigator for the Surveyor television experiments (Whitaker was a member of this team as well), while Kuiper served on the program's Lunar Theory and Processes Working Group. As the Apollo missions gained increasing public attention, the Southwest benefited from greater recognition as well. A few months before *Apollo 11*, *National Geographic Magazine* published an extensive article discussing current knowledge of the Moon and the scientists responsible for that knowledge. The leading figures in this process of discovery included members of the Center of Astrogeology in Flagstaff and faculty at the University of Arizona in Tucson. The region's scientists played other roles during Apollo, coordinating geology field training for astronauts and, in the case of Eugene Shoemaker, chairing the geology teams for *Apollo 11* and *12*. One member of the Flagstaff astrogeology branch enjoyed a rare opportunity to make a dramatic contribution. Harrison Schmitt had joined the Flagstaff facility in 1964, but by the summer of 1966 was well on his way to astronaut status. During *Apollo 17*, the last lunar mission, Schmitt became the only geologist to walk on the Moon.[17]

For all the interest generated by *Ranger, Surveyor*, and *Apollo*, however, the Moon remained only one small part of the solar system. The planets and their satellites had held astronomers in their grip for centuries. Now that technology existed to penetrate space these bodies attracted even greater attention. In the exploration of the planets, southwestern scientists would play an even greater role than they had played in the exploration of the Moon.

The Grand Tour to the Outer Planets: Project Voyager

Kuiper's success in building the Lunar and Planetary Laboratory resulted in increased opportunities for space scientists in the Southwest. In his annual report for 1969–70, for example, Kuiper stressed that the laboratory was involved "in nearly all of the NASA deep-space missions," including those targeted to Mercury, Venus, Mars, and Jupiter. Tom Gehrels, involved with several of these missions, served as a member of the NASA Science Steering Group for Grand Tour. The space agency had begun to design an ambitious mission to the outer planets in 1969, a concept endorsed by a National Academy of Sciences study a few months later. A favorable arrangement of these planets in the late 1970s and early 1980s made it possible to use the planets' gravitational fields to "steer" spacecraft and increase their velocity as well. The "gravity assist" concept would reduce a thirty-year mission to one of twelve years, but only because of a planetary alignment that occurred every 176 years.

Funding realities soon interrupted these ambitious plans. The post-Apollo decline of the American space program, combined with NASA's growing focus on developing the space shuttle, forced a significant scaling down of efforts. Rather than launch four spacecraft on various trajectories to visit Jupiter, Saturn, Uranus, Neptune, and Pluto, NASA scientists redesigned the mission to use only two spacecraft and to visit only Jupiter and Saturn. The downsized program (originally named Mariner Jupiter Saturn, but renamed Voyager in 1977) gained presidential and congressional support in early 1972. The Voyager team would eventually include more than a thousand scientists, engineers, and managers, who would oversee the design and development of nearly a dozen instruments and coordinate mission planning.[18]

The most versatile of the *Voyager* instruments was the imaging system, which would provide a visual record of the probes' encounters with the planets. The leader of the imaging team was Bradford A. Smith, director of planetary programs at New Mexico State University. A protégé of noted astronomer Clyde Tombaugh, Smith had been involved with the development of the astronomy program in Las Cruces as both a student and faculty member since the mid-1950s. He had served as deputy imaging team leader for the *Mariner 9* mission to Mars and participated in several other NASA projects. Especially interested in the Jovian atmosphere, Smith's background in both ground-based astronomy and spacecraft missions provided unique expertise. His deputy team leader on the Voyager mission would be Laurence A. Soderblom, a geophysicist with the Flagstaff astrogeology branch of the U.S. Geological Survey. Following an undergraduate career at New Mexico Institute of Mining and Technology, he pursued his Ph.D. in planetary science and geophysics at California Institute of Technology, where Eugene Shoemaker served as his department chair. Soderblom's focus on satellite geology served as a valuable complement to Smith's interests.

Another important *Voyager* instrument was the Ultraviolet Spectrometer. Of particular value for atmospheric analysis, this experiment would be coordinated by A. Lyle Broadfoot, a native of Saskatchewan who had been a member of the Kitt Peak National Observatory staff since 1963. He had previously directed a similar experiment on the *Mariner 10* mission to Venus and Mercury and represented another example of the growing importance of southwestern scientists to NASA's endeavors.[19]

The role of southwestern institutions in NASA's numerous programs continued to grow in the early 1970s, as indicated by activity at the Lunar and Planetary Laboratory. Director C. P. Sonett noted in his 1973–74 report that laboratory personnel were involved in the analysis of data from Mercury, Venus, Jupiter, and Saturn probes, as well as participating in lunar sample analysis. Many of Sonett's colleagues were also active in design activity for future missions such as Voyager. As the director also stressed in his report, however, the laboratory had urgent need of a senior observational planetary astronomer. Although no formal agreement had been reached as of late spring 1974, the leading candidate for this position was Bradford A. Smith. Not only was he a recognized figure in NASA's interplanetary program but he also had significant experience in ground-based astronomy. He would thus augment the laboratory's existing connection with NASA while providing needed administrative and observational expertise for the Catalina and Mount Lemmon observatories north of Tucson. Smith joined the University of Arizona faculty in August 1974 and soon became a leader in the Tucson space science community.[20]

The launch of the *Voyager* spacecraft in the late summer of 1977 began a twelve-year exploration of the outer solar system. *Voyager 1* reached Jupiter in early March 1979 and returned nearly 19,000 images of the Jovian system. Following a slightly different trajectory, *Voyager 2* approached Jupiter in early July and returned some 14,000 images.[21] From the beginning, the results of the imaging experiments captured the greatest amount of attention. Bradford Smith soon became the best-known member of the Voyager science team because of his active participation in press briefings. His interest in the Jovian atmosphere proved fortuitous, as that topic provided the initial introduction to *Voyager* results. *Voyager 1* observations, both before and during the encounter, showed significant differences from the results of earlier missions. The sophisticated equipment on board *Voyager* revealed much more chaotic atmospheric features and confirmed the concept (based on ground-based telescopic observations) that global variations over time were characteristic of the Jovian atmosphere. The details of circulation patterns and other characteristics revealed by *Voyager*, however, indicated that existing explanations of the planet's atmosphere were inadequate. As Smith remarked in the first daily press briefing on 28 February 1979, "I think, for the most part, we have to say that the existing atmospheric circulation models have all been shot to hell by *Voyager*."[22]

Data supplied by the two *Voyager* spacecraft required significant analysis for a better understanding of the dynamics of the Jovian atmosphere.

Launched in 1977, the two identical *Voyager* spacecraft returned a vast collection of information concerning the outer planets during the 1980s. Southwestern scientists were active members of the teams that designed and operated the imaging system (located at the base of the structure at the top of the photograph) and the ultraviolet spectrometer (located above the circular instrument to the right of the imaging system). (Courtesy of the University of Arizona Space Imagery Center)

Among the most active scientists in this task was Reta Beebe of New Mexico State University. A specialist in planetary atmospheres, Beebe served as director of the Tortugas Mountain Observatory as well as a member of various Voyager teams and groups. Her access to *Voyager* images and her decade-long ground-based observational program allowed her to make significant early contributions to a better understanding of the Jovian atmosphere. Emphasizing that the atmosphere could be characterized by both chaotic features and relatively stable ones, Beebe soon established herself as a leading figure in the study of the atmospheres of the outer planets.[23]

Although the *Voyager* examination of the Jovian atmosphere revealed much of interest, other findings were also dramatic. *Voyager 1* observed no fewer than nine erupting volcanoes on the surface of the Jovian moon Io, eight of which were still active four months later when *Voyager 2* examined the satellite. With plumes rising to altitudes of nearly 200 miles, these volcanoes showed geological activity that astounded planetary geologists and other scientists.[24] The revelation of a thin ring around Jupiter was announced during a press briefing on 7 March and aroused significant interest among the press and public. The photographic sequences for *Voyager 2* were quickly modified to provide additional images making use of better lighting conditions and existing knowledge from *Voyager 1*. These images from July 1979 revealed that Jupiter possessed a system of two main rings. Both rings appeared to be quite thin and composed of small, dark particles, probably similar to dust grains. By no means as dramatic as the rings of Saturn, those of Jupiter provided astronomers with much to consider.[25]

As the Voyager scientists analyzed the Jupiter data, they also began considering the details of the next stage of the mission. Because of the slightly different trajectories of the two spacecraft, *Voyager 1* would reach Saturn in mid-November 1980, to be followed by *Voyager 2* in late August of the following year. Mission planners designed the trajectories of the two *Voyager* spacecraft to return the maximum amount of information. *Voyager 1* would not only provide images and other data concerning the ring system, but would also pass very close to Saturn's largest moon, Titan. This trajectory would deflect the spacecraft's path above the plane of the ecliptic, sending it out of the solar system in that direction. *Voyager 1* could not, therefore, pass close to any of the other outer planets.

Because this spacecraft would examine Titan so closely, however, there existed no need for *Voyager 2* to duplicate these observations. Planners designed the second probe's trajectory to pass through the Saturn system along a path to direct the spacecraft toward Uranus and Neptune, achieving the earlier goal of an exploration of these two distant planets.[26]

As *Voyager 1* approached Saturn during the fall of 1980, the southwestern scientists who had been central to the earlier Jupiter encounter continued to play major roles. Smith and Beebe devoted much time and energy to interpreting the data concerning the Saturnian atmosphere. They emphasized similarities with the dramatic atmospheric conditions on Jupiter, but also stressed that Saturn's greater distance from the Sun and its axial tilt were probably responsible for the differences between the two giant planets. Eugene Shoemaker, a veteran of many NASA programs, was especially interested in the moons of Saturn and found much of value in the images and other data returned to Earth. The Voyager cameras revealed a huge crater on Mimas of some 130 kilometers diameter, indicating an impact that nearly destroyed the moon during the early history of the solar system. The close approach to Titan also revealed that it was not the largest moon in the solar system (as had long been assumed), that the moon's atmosphere was primarily composed of nitrogen, and that various hydrocarbons existed on Titan's surface.[27]

The rings of Saturn, however, proved to be the most dramatic focus of the *Voyager 1* mission. The visual images returned from the spacecraft attracted attention from the national press and led to cover stories in both *Time* and *Newsweek*. Astronomers were even more intrigued. The earlier *Pioneer 11* mission had indicated that the rings were not as simple as they appeared from Earth, but it remained for *Voyager 1* to reveal the incredible complexity of the rings. The existence of several small rings was confirmed during the encounter, but the spacecraft instruments also revealed small satellites called "shepherd moons" that appeared to account for several irregular ring patterns. When the Voyager imaging team met in January to write its mission report, the rings remained the most puzzling aspect of the encounter.[28]

Mission planning for *Voyager 2* emphasized the complementary status of the second spacecraft. The months following the *Voyager 1* encounter allowed scientists to suggest changes in trajectory to provide a closer approach to the rings and to take advantage of better illumination characteristics. The trajectory of the spacecraft was constrained by the gravity assist requirements to send it on to Uranus, but careful planning resulted in more than 18,000 photos and valuable data from other instruments.[29] Atmospheric phenomena were photographed much more effectively, revealing many unusual cloud structures and providing valuable information concerning the dynamics of the planet's weather systems. Saturn's satellites (six of which were discovered by *Voyager*) also attracted much attention. Enceladus, for example, revealed a surface characterized by varied features that suggested a geologically active body. More detailed observations made possible by modifications to trajectory and exposure parameters allowed scientists to examine the ring system with much greater success as well. These observations provided more questions than answers, to be sure, but the revelations concerning Saturn's rings proved to be among the most significant scientific results of the Voyager program.[30]

The discovery of a Jupiter ring system and the revelation of the complexity of the rings of Saturn created a new subdiscipline in space science. When the Voyager mission began in 1977, very little was known about planetary rings, beyond the dramatic image of Saturn as the ringed planet. By the time *Voyager 2* completed its encounter with Saturn during the late summer of 1981, however, several scientists had begun to focus on the ring systems of the two giant planets. Among those who examined these phenomena was Carolyn C. Porco, a graduate student in geological and planetary sciences at California Institute of Technology who served as part of the imaging science team. Saturn data from *Voyager* served as the basis for her doctoral dissertation, which she completed in 1983. One of the most active of the new group of ring theorists, Porco soon joined the Lunar and Planetary Laboratory of the University of Arizona and continued to serve on the Voyager imaging team.[31]

Although the four years between the Saturn and Uranus encounters would provide Voyager scientists with time to analyze existing information, many of them also pursued projects designed to augment the next stage of the mission. Bradford Smith, for example, made several trips to the Las Campanas Observatory in Chile to gain more detailed information about Uranus. This information would be used to plan the *Voyager* flyby more carefully. Equally important, scientists and engineers reprogrammed spacecraft computers to compensate for the low light levels in the outer solar system and for the blurred images caused by *Voyager's* rapid passage through the Uranian system. In addition to reprogramming computers while the spacecraft was in flight, engineers and scientists also modified the spacecraft trajectory so that the cameras could be aimed at the most interesting targets in the Uranian system.[32]

Voyager's encounter with Uranus in late January 1986 produced intriguing results. The spacecraft instruments revealed ten moons in addition to the five previously known. The large satellites appeared to be of a different composition from Saturn's icy large moons, displaying evidence of a greater concentration of rocky material. The rings of Uranus (discovered by telescope in 1977) continued to fascinate Carolyn Porco and her colleagues among the new fraternity of ring theorists. *Voyager* revealed examples of shepherd moons confining the planet's very narrow rings, confirming earlier theoretical explanations of this intriguing phenomenon. Spacecraft cameras also revealed several additional rings around Uranus, all of which were very narrow, relatively close to the planet, and quite dark. The small amount of dust in the rings marked Uranus as quite different from Saturn. Broadfoot's Ultraviolet Spectrometer experiment revealed a bloated hydrogen atmosphere on Uranus, which scientists argued was sweeping the fine dust particles from the rings into the planet.[33]

As *Voyager* completed its encounter at the end of January, its 7,000 images of the Uranian system added much data of value. The impact of these data, however, proved less dramatic than that of the earlier encounters with Jupiter and Saturn. Ironically, the Voyager team's summary of the Uranus results for the press was overshadowed by the space shuttle *Challenger* explosion on 28 January. Even without this space disaster the Uranus results would have lacked dramatic impact. Much less was known about Uranus before the *Voyager* encounter, as compared with Jupiter and Saturn, depriving Voyager scientists of a base from which to compare spacecraft data. As scientists began to contribute their reports and conclusions to the scientific press, Bradford Smith continued to refer to Uranus as "a planet of unfathomable mystery." Smith acknowledged that the uncertainties regarding *Voyager*'s revelations of Uranus were not surprising and the resulting tentative explanations were to be expected. As he told a journalist, however, little possibility existed for the development of definitive explanations due to the lack of any plans for future missions. "We've got to understand Uranus with what *Voyager* gave us," he told Rick

Gore of *National Geographic*. "There's not going to be anything in our professional lifetimes that will tell us more."[34]

Voyager's success at Uranus nonetheless displayed the value of the original "grand tour" concept and convinced NASA administrators to support the last stage of the mission. For relatively little additional cost, the spacecraft could be targeted for a close approach to Neptune during the late summer of 1989. The science and engineering teams could thus be kept intact for the next few years, during which time they would determine how best to modify the spacecraft to gain as much information as possible about the Neptunian system. Despite even greater difficulties than at Uranus, *Voyager* returned more than 9,000 images and provided astronomers with a wealth of data about one of the most mysterious members of the solar system.[35]

Among the topics of great interest to space scientists was the Neptunian atmosphere. From earth-based observations during the 1980s, many of which were made by Bradford Smith in Chile and by numerous other astronomers at Mauna Kea Observatory in Hawaii, it was clear that Neptune possessed intriguing atmospheric features. As *Voyager* began recording Neptune images during the summer of 1989 the planet provided many surprises for Smith, Beebe, and their colleagues. High wind speeds (up to 2,000 kilometers per hour), the persistence of large oval storm systems similar to the Great Red Spot on Jupiter, and the rapid variability of small-scale features all suggested a much more active planet than expected. Spectroscopic measurements revealed that Neptune's deep atmosphere was primarily hydrogen, with approximately 15 percent helium. The planet's blue color came from the small amount of methane in the atmosphere.[36]

Neptune's satellites also attracted much attention. *Voyager* revealed six moons in addition to Triton and Nereid, but the focus of the encounter was the major moon, Triton. Broadfoot's ultraviolet studies disclosed that Triton's atmosphere was thin, transparent, not enveloped by clouds, and much colder than previously thought. Other observations from this instrument revealed that Triton's atmosphere was composed mainly of molecular nitrogen, included a trace of methane near the surface, and was characterized by the presence of auroras. Smith's imaging equipment also revealed much of value. Triton's surface was marked by ridges, cracks, and faults in an icy landscape. A huge ice cap covered most of the southern hemisphere, displaying a blue zone along its edge that suggested fresh frost. This observation indicated a dynamic atmosphere on Triton, as did the near total absence of impact craters on its surface.[37]

Terrestrial observations had revealed rings and partial rings around Neptune, discoveries that astronomers hoped to clarify through *Voyager* imaging. Such observations proved very difficult because of the low light levels at Neptune and the prevalence of dark particles in the rings, but *Voyager* nonetheless provided scientists with much data. Although many of the details of the complete and partial rings (called ring arcs) remained unclear, ring theorists such as Carolyn Porco argued that the best explanation for these curious phenomena involved the same kind of shepherding moons that had been discovered at Saturn and Uranus. More than likely the arcs were portions of complete rings whose invisible sections contained too few particles to be visible. Soon after *Voyager*'s Neptune encounter, Porco established that the newly discovered satellite Galatea was responsible for the confinement of some of the ring arcs.[38]

Voyager's encounter with Neptune marked the end of one of the most extensive planetary reconnaissances of all time. At a cost of some $865 million, the two spacecraft returned nearly 115,000 images and large quantities of other data from the many experiments on board. For more than a decade, scientists and engineers had guided the spacecraft and analyzed the information that came back to Earth. *Voyager 2* headed into deep space in the late summer of 1989 on a different path from its twin, which had begun its journey out of the solar system nearly nine years earlier.[39]

The individuals involved with the Voyager program pursued different paths as well. By the spring of 1990 Bradford Smith had retired from the University of Arizona to his home in Hawaii, where he would continue his work on planetary science. Reta Beebe returned to her position at New Mexico State University where she and her colleagues continued to study the Jovian atmosphere. Data from both the *Hubble Space Telescope* and the *Galileo* spacecraft (which went into orbit around Jupiter in December 1995) allowed Beebe to continue her work throughout the 1990s. Carolyn Porco, now a senior researcher at the Lunar and Planetary Laboratory, returned to Tucson to analyze in greater depth the *Voyager* data concerning planetary rings. Her discussion of the shepherding moon concept became a central tenet of explanations of rings and ring arcs. Porco's status in the profession involved her in various NASA projects and resulted in her selection as imaging team leader for the *Cassini* spacecraft scheduled to reach Saturn in 2004.[40]

The Voyager missions supplied scientists with significant quantities of data and revealed the outer planets of the solar system as dramatically intriguing astronomical objects. Despite NASA's concentration on other programs such as the space shuttle and the resulting lack of funds for the agency's science efforts, Voyager results proved among the most important scientific advances of the 1980s. The Voyager missions, however, represented the end of a chapter in NASA's planetary exploration. Although a few large missions remained, the goal of NASA during the 1990s was increasingly the use of small, inexpensive spacecraft to provide data about the solar system. This new vision of space exploration would require significant rethinking of how to design and build the required instruments. Once again, scientists in the American Southwest would be involved with much of this activity.

Return to the Red Planet: Mars Pathfinder

During the summer of 1997 people around the world gazed at their television screens or computer monitors to see the surface of Mars for the first time in two decades. Not only did they witness the panorama of images beamed to Earth, but they also watched as a small rover vehicle named "Sojourner" crept across the Martian landscape. Scientists were as fascinated by the data returned from *Mars Pathfinder* as the public was intrigued by the images. As had been the case with most earlier planetary probes, many of these scientists were associated with institutions in the American Southwest.

Pathfinder's origins began five years earlier as the first example of a new perspective on spaceflight. Appointed head of NASA in 1992 after a career in the aerospace industry, Daniel Goldin brought a devotion to turning the space agency into an organization more in line with his corporate experience. To that end, he proposed smaller and less costly missions, a redistribution of funds toward the development of new technologies, and the immediate dissemination of information from missions on the World Wide Web. The days of large spacecraft designed to accomplish as much as possible in one mission were over. In their place, NASA would support small spacecraft designed and at least partly assembled by small groups of scientists and engineers at universities and in industry. Mars Pathfinder, approved by NASA in 1992, had a budget ceiling of $150 million, approximately 5 percent of the cost for the Viking missions of the mid-1970s. The new technology focus was also evident in early planning for Pathfinder, as NASA funded a preliminary experiment to develop a Mars surface vehicle to be part of the mission. This device became the Microrover Flight Experiment later named "Sojourner."[41]

As a result of the "faster, better, cheaper" focus, Pathfinder designers were faced with numerous cost-cutting decisions and were forced to simplify wherever possible. Decisions concerning the landing of the spacecraft displayed these considerations dramatically. To save money, *Pathfinder* would travel directly to Mars; there would be no orbiter/lander combination as in *Viking*. Descending through the Martian atmosphere, the spacecraft would be slowed by parachute and small retro-rockets, but its primary protection would be a set of pressurized airbags to cushion its impact on the surface. Estimates suggested that the airbag-surrounded spacecraft would hit the surface and bounce several times before coming to rest. The bags would then deflate and the mission could begin. As the design of this system progressed through the end of 1995, engineers relied on computer models, drop tests, and other activities to gauge more accurately how the system would function. Despite these and other complications with the design of the spacecraft, Pathfinder remained a relatively small NASA program with no more than fifty engineers and a hundred scientists on the program at any one time.[42]

Many of these scientists and engineers were involved with the instruments that would examine the Martian surface. As had been the case with the *Voyager* probes of the previous decade, the imaging equipment proved to be the most versatile. Once again University of Arizona personnel played a leading role in this part of the mission. The new NASA focus, however, resulted in an intriguing change. Not only would the Imager for Mars Pathfinder (IMP) be designed at the Lunar and Planetary Laboratory, it would also be assembled and tested in Tucson before being shipped to Pasadena. Peter H. Smith, the team leader for the camera system, later estimated that the six million dollar cost of the equipment would have been tens of millions of dollars had the old methods been followed. By the end of December 1995, following more than a week of frantic final assembly activity, IMP was shipped to the Jet Propulsion Laboratory for final testing and mating with the spacecraft. The Tucson scientists had tested their equipment throughout the year in a "Mars Garden" built beside the Kuiper Space Sciences Building on campus and were confident that all would work as designed. Less than a year later, Smith and a dozen other University of Arizona scientists watched the launch of *Mars Pathfinder* from Cape Canaveral.[43]

Although the visual images from the surface of Mars became the best-known results of the IMP camera, the equipment was designed to achieve much more. The stereoscopic camera, mounted on a three-foot high mast, would return three-dimensional images to provide a better understanding of the topography of the landing site. One of its first tasks after landing would be to find the Sun at sunrise. This was necessary so that the lander computer could aim the high-gain antenna (the high-speed communications link) toward Earth. The landing technique made it impossible to know precisely where the spacecraft came to rest, so aiming the antenna could only be done with surface equipment. The camera would also be necessary to survey the lander to determine which ramp would be the best exit for the rover. The camera would then confirm that the rover had deployed properly and would be essential for the engineers who were "driving" this robot on the Martian surface. Several members of the IMP team planned to use the camera for specific scientific observations as well. Daniel T. Britt of the Lunar and Planetary Laboratory hoped to identify surface rocks and minerals, including some that would be good targets for rover instruments. Ronald Greeley and Robert Sullivan of Arizona State University would coordinate a wind sock experiment to determine surface wind speeds and direction, while relying on other IMP experiments to measure the size and density of atmospheric dust particles. Long-time NASA participant Laurence Soderblom of the U.S. Geological Survey branch in Flagstaff planned to rely on stereoscopic images to study the geology of the landing area from a unique perspective.[44]

As final testing of the spacecraft took place in Pasadena and Cape Canaveral during the summer and fall of 1996, public interest in the mission was heightened by a dramatic announcement by NASA scientists.

In early August a team led by geologist David McKay of the Johnson Space Center in Houston reported the discovery of complex, relatively large organic molecules in a meteorite of Martian origin. These scientists also revealed that under an electron microscope, certain small oval shapes appeared that were similar to microfossils on Earth, although nearly one hundred times smaller in size. McKay and his colleagues explained their results as evidence of the existence of primitive life in the early history of Mars, creating yet another public fascination with the prospect of life on the red planet.

This interest quickly found expression in the nation's capital. Vice President Albert Gore and Speaker of the House Newt Gingrich assured NASA administrator Goldin that additional funds for Mars exploration would likely be available. President Bill Clinton announced a White House conference on space exploration, telling reporters, "I am determined that the American space program will put its full intellectual power and technological prowess behind the search for further evidence of life on Mars." Goldin and other NASA officials attempted to stress in their comments that "good science" would have to remain the space agency's priority, but the public fascination with the possibility of life on Mars could not be ignored as a source of additional support for NASA and its budget. Although scientists soon developed more likely interpretations of the meteorite evidence that had nothing to do with life, the public's interest in *Mars Pathfinder* never wavered.[45]

Mars Pathfinder successfully landed on the Martian surface on 4 July 1997. Peter Smith and his IMP colleagues waited anxiously for data from the high-gain antenna to show that the camera had located the Sun, but were soon busy coordinating other activities. Slightly more than six hours after landing the camera began taking a frame-by-frame panorama of the landing site from its stowed position. These images served as insurance in case any malfunctions prevented the complete deployment of the IMP equipment and provided the dramatic focus of the news conference that began at 6:30 p.m. in the JPL auditorium. Smith revealed the image to reporters, after which it appeared on television newscasts and the front pages of newspapers around the world. The IMP was fully deployed the following day, allowing the camera to take a full 360-degree panorama that made complete use of the instrument's three-dimensional capabilities.

The images and other data that *Pathfinder* returned to Earth provided scientists and the public with an unparalleled vision of the Martian surface. Even before the rover began its exploration of the landing site the images from IMP provided valuable insight into the geology of the area. Smith captured the feeling of many when he observed during the 6 July news conference that "These last few days have been the greatest thrill of my professional life." IMP images were soon augmented by other images from the rover cameras as "Sojourner" explored nearby rocks. Although the public was intrigued by images of rocks that became known by names such as "Yogi," "Barnacle Bill," and "Bamm-Bamm," scientists were deluged with numerous other sources of data from the several instruments on board the lander and the rover.[46]

The latest Mars mission continued to attract attention throughout the summer. Two weeks after the successful landing, Smith and five other Pathfinder scientists accompanied Goldin to meetings with congressional leaders and with President Clinton. The public's interest in the mission was further shown by the high traffic at the various sites on the World Wide Web. Goldin's goal of making the results of NASA missions widely available found great support among the public. Between early July and early August, Pathfinder sites received more than 560 million "hits" from throughout the world. Greatly exceeding design expectations, *Pathfinder* continued to provide data until late September. The lander and the rover sent back nearly 17,000 image frames, as well as data from other instruments. Both the public and the scientific community had received far more information than anticipated from the small spacecraft.[47]

By the end of the year, Pathfinder scientists had analyzed much of the data from the mission and had begun to disseminate their results and conclusions. The noted periodical *Science*, the official journal of the American Association for the Advancement of Science, devoted much of its 5 December issue to the mission. Articles discussing the results of the various experiments were included as were several of the most dramatic images from the *Pathfinder* cameras. Scientists, many of whom were associated with southwestern institutions, reported on such diverse topics as surface meteorology and climate, atmospheric characteristics, and the magnetic properties of particles suspended in the Martian atmosphere.[48]

Although many instruments were involved in these experiments, the IMP cameras proved especially useful. Images returned from this equipment included surface features and rocks, lander images, details of atmospheric properties, and astronomical observations. Shortly after IMP had been fully deployed, team member Nick Thomas of the Max Planck Institute for Aeronomy in Germany took the first of several images of the Martian moon Deimos. Because of its close proximity to the planet this satellite had always been difficult to observe from Earth or from the *Hubble Space Telescope*. IMP images revealed valuable data concerning the spectrum and composition of Deimos, especially when compared with similar data from the earlier Viking mission. Images of the sky above the landing site were used to determine the nature of aerosol particles in the atmosphere, providing information on solar energy and other climatic phenomena. Observations of the Sun and sky through various filters allowed IMP team members to determine the amount of water vapor in the Martian atmosphere. An intriguing use of the IMP camera involved images of the wind sock experiment on the lander. As a result of regular observations of the three wind socks deployed at the beginning of the mission, Arizona State University scientists Greeley and Sullivan were able to determine that wind velocities were generally less than eight meters per second during the Pathfinder mission.[49]

The topic of greatest interest to the general scientific community was the geology of Mars. In the observations and experiments investigating this topic, the IMP and the rover achieved something of a symbiotic relationship. Images from the IMP and rover cameras were combined so that "Sojourner" could be driven to locations of interest. The complementary nature of IMP and the rover was particularly useful in the analysis of Martian rocks. Direct measurements of rock composition by rover instruments were paralleled by IMP observations of subtle color variations in these rocks, leading geologists to conclude that the rocks in the landing site area were quite similar to each other.[50] Rover images further suggested the possibility that some of the surface rocks were sedimentary in origin, but this conclusion was increasingly challenged in the following months. When the American Geophysical Union met in San Francisco in December, the consensus was that the rocks of the landing site were volcanic in origin. Nonetheless, *Pathfinder* results confirmed the presence of surface water in the Martian past and led scientists to conclude that the Martian climate was both warmer and wetter at some point in the planet's history. A significant climatic change had clearly taken place, but the information necessary to estimate when that change had taken place would require future Mars missions.[51]

As the results of the Pathfinder mission continued to attract attention during the late 1990s the importance of scientists from the Southwest remained evident. The Voyager missions of the previous decade had certainly demonstrated the growing role of the region's scientists in the American space program, but Pathfinder indicated that southwestern institutions were now playing an even greater role. This situation was confirmed as other missions began reporting results. The *Mars Global Surveyor*, which entered Mars orbit just as *Pathfinder* was concluding its observations, included scientists from the University of Arizona, Arizona State University, and the Flagstaff branch of the U.S. Geological Survey.[52] *Lunar Prospector*, a desk-sized spacecraft that cost a mere $63 million, entered polar orbit around the Moon in January 1998. The first NASA-supported lunar mission since Apollo, *Lunar Prospector* was designed to map the Moon from various altitudes before crashing into the surface during the summer of 1999. Experiments to determine magnetic and gravitational properties involved Lon L. Hood of the University of Arizona, but the most noteworthy results of the mission came from the neutron spectrometer experiment coordinated by scientists at the Los Alamos National Laboratory. This instrument revealed evidence of frozen water near the lunar poles and served as yet another example of the results that could be obtained from smaller and cheaper spacecraft.[53] As NASA designed and planned additional planetary missions, the agency continued to rely on the expertise of southwestern scientists.

The roles played by southwestern scientists in the Ranger, Voyager, and Mars Pathfinder missions serve as an intriguing overview of the evolution of the region's status as a scientific center. The work of Gerard Kuiper in the foundation of the region's involvement with the American space program was immediately successful. By the time of the successful flight of *Apollo 11* in 1969, Kuiper's Lunar and Planetary Laboratory was playing a major role in several NASA planetary programs. Other institutions in the Southwest, such as New Mexico State University and the Flagstaff branch of the U.S. Geological Survey, were increasingly evident in NASA's programs as well. The Voyager mission to the outer planets revealed how firmly entrenched the region's scientists had become in interplanetary exploration, a status that enabled the expansion of space science programs in the Southwest. As the twenty-first century opened, the major universities and other science institutions in the Southwest remained actively involved in the exploration of the solar system, building on the foundations laid more than forty years earlier.

Chapter 10 From Natural Laboratory to Scientific Center

The topics discussed in this volume reveal the diversity of scientific activity in the American Southwest and contribute to a more complete portrait of the region's past. Science has long been ignored in historical accounts, but its role in the development of the Southwest was significant. An examination of the region's science also provides a fuller account of the growth of science in the United States, as the Southwest has usually been overlooked in historical treatments of the broader topic.

The concept of "regional" science is, to be sure, a problematic one. Physical laws do not change as one travels from place to place but the practice of science frequently displays certain unique regional characteristics. In the American Southwest these characteristics provide themes that help to articulate the science practiced in the region. These themes also guide the historian in the effort to organize disparate material into a more coherent perspective on the region's scientific development.

From the earliest European penetration the concept of the Southwest as a natural laboratory played a central role. Explorers from Spain, Mexico, and the United States surveyed the Southwest from the sixteenth into the nineteenth century as part of a general reconnaissance of the area. This interest continued during the late nineteenth century as large numbers of naturalists surveyed the area for museums and other institutions. Such explorations continued well into the twentieth century, providing scientists with new information in botany, zoology, climatology, and other specialties. The setting of the Southwest was important to other disciplines as well. Archaeology and anthropology benefited from the well-preserved ancient ruins, while astronomy profited from access to the clear and steady atmosphere of the region, increasingly important to the sophisticated research of the twentieth century.

The Southwest also displayed a "colonial" characteristic in its scientific development. Because of its sparse population the region's institutional base remained limited for decades. Science and other intellectual activities were imported from other regions of the nation, as were many social and political institutions. Most scientists active in the region had been born and educated elsewhere; many came from eastern states, although California also provided part of this population. These individuals participated in the surveys mentioned above and were employed by southwestern universities, experiment stations, and other institutions as they expanded during the first half of the century. Only later did the region's educational and scientific facilities begin to play a role in supplying the members of this population. Increasingly through the early twentieth century a scientific community emerged in the region.

The research agendas of scientists in the Southwest also displayed a colonial perspective. Moving into the region, these scientists focused their training and experience on the Southwest's unique characteristics. One of the most dramatic examples of colonial activity was the establishment and growth of the research station at Los Alamos. Initially located in northern New Mexico because of its isolation, the laboratory soon became the major focus of the region's science. The situation was similar for applied science, as large numbers of eastern scientists and engineers played major roles in the mining industry. Agriculture also had an increasingly scientific base beginning in the late nineteenth century. Regional institutions such as the agricultural experiment stations hired staff members from eastern and midwestern universities to pursue projects designed to place commercial agriculture on a more rational and more profitable basis.

The growing importance of scientific institutions serves as another theme that defines science in the American Southwest. Especially during the nineteenth and twentieth centuries, scientific institutions became the central aspect of the growth of many disciplines throughout the United States. Because of its low population, the Southwest joined this process relatively late. The institutional base of the region's science began in the late nineteenth century with the establishment of the territorial universities and agricultural experiment stations. Another important early example of such growth was the Lowell Observatory, founded in 1894, but more significant expansion was a product of the early twentieth century. The Carnegie Desert Botanical Laboratory, Steward Observatory, Museum of Northern Arizona, and the institutions that would become the Museum of New Mexico and the Laboratory of Anthropology provided an important base for the region's scientists. Federal support trickled into the region at the same time through the expansion of U.S. Department of Agriculture facilities to pursue topics in commercial agriculture, range management, and forestry.

The Department of Agriculture's presence in the Southwest reflects the importance of the federal government in the region's scientific development. This theme is most clearly shown by the development of Los Alamos and related facilities during and immediately after World War II, but even before the war the federal government played a central role. The Department of Agriculture was instrumental in the establishment of the agricultural experiment stations and provided important funding for the territorial universities. The department also established its own institutions in the region to pursue numerous research programs. The Department of the Interior supported archaeological investigations throughout the region. Such federal programs became especially notable during the New Deal, as the Roosevelt administration expanded various programs and provided other funding through relief agencies. These efforts frequently provided money to universities to hire students and other employees, many of whom were assigned to science departments and research projects.

The expansion of federal funding that followed World War II also had a significant impact on the growth of science in the Southwest. Even with its modest institutional base, the region enjoyed the benefits of federal money. Beginning in the late 1950s, for example, astronomy and space science were especially well supported, as indicated by the selection of Kitt Peak in southern Arizona as the national observatory site. The National Science Foundation provided much of the funding for this project, as it did for numerous astronomy programs in the colleges and universities of Arizona and New Mexico. The National Aeronautics and Space Administration provided generous support for several space science programs that quickly established the Southwest as a center for space research. As was the case throughout the United States, federal funding flowed into the region's colleges and universities to support basic research and science education. The growth brought about by this federal funding played an important role in establishing the Southwest as a partner in the national scientific establishment, a position it continued to hold in the early years of the twenty-first century.

The historical analysis of the development of science in the American Southwest provides many opportunities for further study. The themes of "natural laboratory" and "colonial science" suggest several topics. Although the work of explorers in the region during the mid-nineteenth century has been described, the impact of their collections on eastern museums and the role such collections played in the development of science remain unexamined. Equally important, the large numbers of naturalists who explored the Southwest in the late nineteenth and early twentieth century have received only intermittent attention. The numerous publications contributed by these naturalists indicate a significant amount of activity for several decades. Freelance naturalists, museum staff, and university faculty all appear to have participated, suggesting that a large amount of material remains to be located and analyzed. The development of medicine in the Southwest could be similarly investigated. Historians have established that the region served as a haven for health-seekers, especially those plagued with respiratory ailments, but a more detailed examination of the treatment these patients received would provide insight concerning the region's scientific development. Institutional factors undoubtedly played an important role, suggesting that an analysis of the various hospitals and sanatoria in the region would provide much information of interest. Similarly, the impact of this aspect of the health industry on local culture and society would interest historians concerned with the social history of the region in the late nineteenth and early twentieth century.

Institutional developments and the role of the federal government in the Southwest also offer great promise as historical research topics. Central to the growth of science was the establishment and expansion of the region's universities and colleges. The role of science and the science faculty in these institutions is not well known, especially in the territorial and early statehood periods. Although the analysis of those southwestern scientists listed in the various *American Men of Science* volumes provides an understanding of the characteristics of the region's leading scientists, these individuals represent only a portion of the scientific population. Most universities in the region possess archival material that would provide the basic information about these individuals. An analysis of this material would yield a valuable portrait of the specialties and backgrounds of the region's academic scientists. An examination of the rapid growth of universities following World War II would highlight the changing status of science in higher education and the increasing importance of federal support for such activity. Not only did the postwar enrollment boom provide universities with more money from tuition and state appropriations, but the recognition of the importance of science in the curriculum disproportionately provided science departments and programs with generous budgets.

Developments in applied science offer significant potential for historical analysis, especially in agricultural science. A central aspect of the region's economy throughout its history, agriculture became increasingly involved with scientific research beginning in the late nineteenth century. The U.S. Department of Agriculture provided funding for the region's universities and experiment stations while expanding its own laboratories and other research facilities. The Forest Service, for example, established the nation's first forest experiment station near Flagstaff in 1908, followed by the creation of range preserves in southern Arizona and southern New Mexico in the late 1920s. The Southwest Forest and Range Experiment Station was established in Tucson in 1930 to coordinate the various research programs in forestry and range management. Details of these research programs, as well as of the institutional development of these facilities, would provide significant knowledge concerning the role of science in the region's use of its natural resources.

Significant work was also performed by the region's universities and agricultural experiment stations. Nutritionists joined home economics faculties and experiment station staffs, collaborating with agricultural scientists in various projects. This combination of backgrounds and specialties characterized much experiment station activity, establishing an interdisciplinary tradition in the region's agricultural research. Staff members sought methods to improve existing commercial agriculture and searched for promising new crops. In New Mexico, much of the experiment station's most famous research focused on establishing chile as a leading aspect of the state's economy, but the Las Cruces facility also investigated the commercial value of fruits and vegetables, creating a more diversified agriculture. Arizona's experiment station scientists concentrated on expanding the state's citrus industry, improving the use of the range for livestock, and further establishing cotton as a major crop. Here, too, research concerning other fruits and vegetables played an important role in experiment station research. Detailed historical investigations of such research would provide a better grasp of the important links between science and agriculture in the American Southwest.

The growth of the American Southwest from its territorial beginnings to its current place as a focus of the Sun Belt has involved many events and ideas. Population growth, economic expansion, and political maturation all played crucial roles in this development and have received attention by historians eager to portray the region in all its complexity. Much less attention has been paid to the development of science in the region, but this phenomenon also contributed to the Southwest's development. Especially after World War II, science played an important part in the life of the United States. The Southwest shared in this postwar phenomenon, but the region had already established a recognizable scientific foundation by the 1930s. This foundation made possible the region's later emergence as a leading center of science. An historical exploration of the development of science in the Southwest thus provides a more complete perspective on the region's growth and development. The region's unique physical characteristics early defined the role science played in the Southwest, a role expanded, but never supplanted, by later developments. While offering additional insight concerning the integration of the region into the United States, the history of science also reminds us that the Southwest remains a unique part of the nation.

Notes

Introduction

1. Nash, *The American West in the Twentieth Century*, 127–129, 206–209, 258–261, 297; Nash, *The American West Transformed*, 153–177. An excellent example of a specialized study is Szasz, *The Day the Sun Rose Twice*.

2. Such perspectives are shown by several of the essays in Fernlund, *The Cold War American West, 1945–1989*, and by Ackland, *Making a Real Killing*. More balanced treatments include Gerber, *On the Home Front*, and Hevly and Findlay, *The Atomic West*.

3. The classic statement of the "New Western History" remains Limerick, *The Legacy of Conquest*, but also see White, *"It's Your Misfortune and None of My Own"* and Milner, *The Oxford History of the American West*. A guide to the scholarly debate concerning this new vision of the American West is Limerick, *Trails: Toward a New Western History*.

4. In addition to Szasz's study of the Trinity test, see Hoddeson, *Critical Assembly*, and Furman, *Sandia National Laboratories*.

5. Hoyt, *Lowell and Mars*; Hoyt, *Planets X and Pluto*; Webb, *Tree Rings and Telescopes*; Strauss, *Percival Lowell*; Evans and Mulholland, *Big and Bright*; Edmondson, *AURA and Its US National Observatories*.

6. Haury, *Point of Pines, Arizona*; Parezo, *Hidden Scholars*; Woodbury, *Sixty Years of Southwestern Archaeology*; Reid and Whittlesey, *Archaeology of Ancient Arizona*; Elliott, *Great Excavations*; Fowler, *A Laboratory for Anthropology*.

7. Examples of such studies include Colley, *The Century of Robert H. Forbes*; Spidle,

Doctors of Medicine in New Mexico; Hoyt, *Coon Mountain Controversies*; and Bowers, *A Sense of Place*.

Part I. Introduction

1. Lanham, *The Bone Hunters*, 126–137; "Progress of the Ungulates in Tertiary Time," 1055–1057; "The Oldest Tertiary Mammalia," 385–387. Cope's major publica tions concerning his New Mexico research include "On some Mammalia of the Lowest Eocene beds of New Mexico," 484–495; "Synopsis of the Vertebrata of the Puerco Eocene Epoch," 461–471; "First Addition to the Fauna of the Puerco Eocene," 545–563; "Second Addition to the Knowledge of the Puerco Epoch," 309–324; "The Permian Formation of New Mexico," 1020–1021; and "Synopsis of the Vertebrata Fauna of the Puerco Series," 298–361.

2. The museum's investigations of mammalian fossils are reported in Osborn and Earle, "Fossil Mammals of the Puerco Beds. Collection of 1892," 1–70; Matthew, "A Revision of the Puerco Fauna," 259–323; and Sinclair and Granger, "Paleocene Deposits of the San Juan Basin, New Mexico," 297–316. For a summary of research concerning the region's dinosaur fossils, see Brown, "The Cretaceous Ojo Alamo Beds of New Mexico with Description of the New Dinosaur Genus *Kritosaurus*," 267–274.

3. An excellent survey of the role played by scientists and engineers in western mining is Spence, *Mining Engineers in the American West*.

4. Ibid., 44, 126, 212–213, 249, 275–276, 342; Spence, "The Janin Brothers: Mining Engineers," 76–82.

5. Webb, "Scientists in the American Southwest," 176–177, 179–180. A detailed discussion of New Mexico School of Mines is Christiansen, *College on the Rio Grande*.

6. Webb, "Scientists in the American Southwest," 176–177; Rice, "Arizona Agricultural Experiment Station," 123–140. The role of scientists in experiment stations is chronicled in Rosenberg, *No Other Gods*, 153–199.

7. Rice, "Arizona Agricultural Experiment Station," 131–132; Toumey, "The Date Palm," 102–150; Swingle, "The Date Palm and Its Utilization in the Southwestern States," 11–155.

8. Rice, "Arizona Agricultural Experiment Station," 132; Vinson, "The Stimulation of Premature Ripening by Chemical Means," 208–212; Vinson, "Chemistry and Ripening of the Date," 403–435; Freeman, "Ripening Dates by Incubation," 437–456; Colley, "Arizona, Cradle of the American Date Growing Industry, 1890–1916," 61–64.

9. Webb, "Scientists in the American Southwest," 178; Rice, "Arizona Agricultural Experiment Station," 140.

10. On the Lowell Observatory, see Hoyt, *Lowell and Mars*; Hoyt, *Planets X and Pluto*; and Putnam, *The Explorers of Mars Hill*.

11. Discussions of the development of southwestern archaeology include Elliott, *Great Excavations*, and Reid and Whittlesey, *Archaeology of Ancient Arizona*. Also see Rothman, *Preserving Different Pasts*, 6–88.

12. Webb, "Scientists in the American Southwest," 177–178; McGinnies, *Discovering the Desert*, 1–16.

13. On Douglass and dendrochronology, see Webb, *Tree Rings and Telescopes*; Webb, "Solar Physics and the Origins of Dendrochronology," 291–301; and Nash, *Time, Trees, and Prehistory*.

Chapter 1. The Desert Revealed

1. Nabhan, *Gathering the Desert*, 15; Wilson, "Mesquite," 97–100; Kunitz, *Disease Change and the Role of Medicine*, 140–141.

2. Palmer, "Plants Used by the Indians of the United States," 595–596; Nabhan, *Gathering the Desert*, 137–149.

3. Hurt, *Indian Agriculture in America*, 16–26; Reid and Whittlesey, *Archaeology of Ancient Arizona*, 214–215, 227; "Arizona and its Heritage," 199–203; Loftin, *Religion and Hopi Life in the Twentieth Century*, 4–12.

4. Reid and Whittlesey, *Archaeology of Ancient Arizona*, 69–110, 222–223, 244–251; Haury, *The Hohokam*, 120–151.

5. Hurt, *Indian Agriculture in America*, 42–57.

6. McCluskey, "Historical Archaeoastronomy," 31–57; Colton, *Hopi Kachina Dolls*, 2–3; Fewkes, *Hopi Katcinas*, 22, 52, 89; Waters, *Book of the Hopi*, 168–187. For an overview of archaeoastronomy in the region, see Malville and Putnam, *Prehistoric Astronomy in the Southwest*.

7. Williamson, "Casa Rinconada," 205, 213–216.

8. Sofaer, "Lunar Markings on Fajada Butte," 169–181; Sofaer, "Unique Solar Marking Construct," 283–291.

9. Williamson, "Anasazi Solar Observatories," 203–204, 211–212.

10. Brandt, "Possible Rock Art Records," 46–56; Brandt and Williamson, "Rock Art Representations," 171–176; Ellis, "A Thousand Years," 60–65, 71–74, 84–87.

11. Bolton, *Coronado*, 131, 188. For details of the early Spanish efforts, see Reinhartz and Jones, "*Hacia el Norte!*" and Jones, "Spanish Penetrations." For the broader context of Spanish exploration in North America, see Weber, *The Spanish Frontier in North America*.

12. Jones, "Spanish Penetrations," 54–55; John L. Kessell, "To See Such Marvels," 69, 73.

13. The standard Kino biography remains Bolton, *Rim of Christendom*. On Kino's explorations in the Southwest, see pp. 59–60, 75, 92–104, 170–205, 271–286, 369–486, 550–555, 565–570, 592. His description of Casa Grande appears on p. 285.

14. Kessell, "To See Such Marvels," 69–71, 74; Jones, "Spanish Penetrations," 39–40, 49–52, 60–62; Engstrand, *Spanish Scientists in the New World*, xi, 3–11, 184–185. On the Royal Corps of Engineers, see Fireman, *The Spanish Royal Corps of Engineers in the Western Borderlands*.

15. Goetzmann, *Exploration and Empire*, 43–51.

16. Goetzmann, *Army Exploration*, 4–5, 12–17, 59–61.

17. Ibid., 127–147; Norris, *Emory*, 35–64.

18. Goetzmann, *Army Exploration*, 153–208; Norris, *Emory*, 65–172; Emory, "Running the Line," 221–265. For a detailed discussion of the Mexican commission, see Hewitt, "The Mexican Boundary Survey Team," 171–196.

19. Emory, "Running the Line," 240–243; Norris, *Emory*, 156; Goetzmann, *Army Exploration*, 197–204.

20. Norris, *Emory*, 163–172; Goetzmann, *Army Exploration*, 204–205.

21. Goetzmann, *Army Exploration*, 209–212, 262–275, 306–311. An excellent overview of science during this period is Bruce, *The Launching of Modern American Science, 1846–1876*.

22. Conrad, "The Whipple Expedition in Arizona, 1853–1854," 149–151, 161, 170–171, 175–176; Gordon, *Through Indian Country to California*, 174; Goetzmann, *Army Exploration*, 287–292.

23. Goetzmann, *Army Exploration*, 305–337.

24. Ibid., 378–397.

25. Bartlett, "Scientific Exploration of the American West, 1865–1900," 461–465; Goetzmann, *Exploration and Empire*, 391–393. A comprehensive overview of post–Civil War activity is Bartlett, *Great Surveys of the American West*.

26. Goetzmann, *Exploration and Empire*, 530–576; Bartlett, *Great Surveys*, 219–329; Powell, *Exploration of the Colorado River*, 22, 32. The standard biographies of Powell are Stegner, *Beyond the Hundredth Meridian* and Worster, *A River Running West*.

27. Goetzmann, *Exploration and Empire*, 467–475, 483–488, 592–601.

28. Cutright and Brodhead, *Elliott Coues: Naturalist and Frontier Historian*, 45–46, 51–90.

29. Ibid., 65–68, 85–87, 91–92, 178–184. Coues, "Ornithology of a Prairie-Journey," 157–165; Coues, "Notes on various birds observed at Fort Whipple, Ariz.," 535–538; Coues, "List of the Birds of Fort Whipple, Arizona," 39–100; Coues, *Birds of the Colorado Valley*.

30. Cope, "On the REPTILIA and BATRACHIA of the Sonoran Province of the Nearctic Region," 300–314; Coues, "The Quadrupeds of Arizona," quoting p. 281; Coues, "Notes on a Collection of Mammals from Arizona," 133–136.

31. Richmond, "In Memorium: Edgar Alexander Mearns," 1–18.

32. Mearns, "Description of a rare Squirrel, new to the Territory of Arizona," 297–307; Mearns, "Description of supposed New Species and Subspecies of Mammals, from Arizona," 277–307; Mearns, "Description of a New Species of Weasel, and a New Subspecies of the Gray Fox, from Arizona," 234–238; Mearns, "Notes on the Otter (Lutra canadensis) and Skunks (Genera Spilogale and Mephitis) of Arizona," 252–262.

33. Mearns, "Descriptions of a New Species and Three New Subspecies of Birds from Arizona," 243–251; Mearns, "Observations on the Avifauna of Portions of Arizona," 45–55, 251–264.

34. Richmond, "Mearns," 8–9; Mearns, "Descriptions of Three New Forms of Pocket-Mice from the Mexican Border of the United States," 299–302.

35. Jepson, "Lemmon, John Gill," 162–163; Crosswhite, "J. G. Lemmon & Wife," 12–14, 17–18; Bonta, *Women in the Field*, 74.

36. Crosswhite, "J. G. Lemmon & Wife," 14–15, 18–21; Bonta, *Women in the Field*, 74–75. Lemmon, "A Botanical Wedding Trip," 517–525; Lemmon, "Botanizing in Arizona," *Mining and Scientific Press*, 18 June 1881.

37. Crosswhite, "J. G. Lemmon & Wife," 15–21.

38. "Notes and News," *Auk* 30 (July 1913): 472; Brown, "Bendire's Thrasher," 225–231; Brown, "A Band-tailed Hawk's Nest," 392–393; Brown, "Arizona Bird Notes," 43–50; Brown, "Masked Bob-white (*Colinus Ridgwayi*)," 209–213; Allen, "The Masked Bobwhite (Colinus Ridgwayi) of Arizona, and its Allies," 273–290.

39. James, "Botanical Notes from Tucson," 978–987.

40. "Merriam's 'Results of a Biological Survey of the San Francisco Mountain Region,'" 95–98.

41. Toumey, "A bit of the flora of Central Arizona," 162–164; Cockerell, "Notes on New Mexican Flowers and Their Insect Visitors," 104–107.

42. Allen, "On a Collection of Mammals from Arizona and Mexico," 193–258.

43. Minckley, "Frederic Morton Chamberlain's 1904 Survey of Arizona Fishes," 177–237.

44. Kofalk, *No Woman Tenderfoot*, 27–28, 44–46, 48, 54–73, 86–88; Bonta, *Women in the Field*, 186–191.

45. Bonta, *Women in the Field*, 193; Kofalk, *No Woman Tenderfoot*, 92, 101–116. Bailey, "Additional Notes on the Birds of the Upper Pecos," 349–363; Bailey, "Additions to Mitchell's List of the Summer Birds of San Miguel County, New Mexico," 443–449; Bailey, "A Drop of Four Thousand Feet," 219–225.

46. Bailey, "Notable Migrants Not Seen at Our Arizona Bird Table," 397–398, 408–409.

47. Kofalk, *No Woman Tenderfoot*, 162–169.

Chapter 2. Applying Science

1. Discussions of mining promoters and consultants include Atherton, "The Mining Promoter in the Trans-Mississippi West," 35–50; Hinton, "Frontier Speculation: A Study of the Walker Mining Districts," 245–255; Paul Lucier, "Scientists and Swindlers: Coal, Oil, and Scientific Consulting in the American Industrial Revolution, 1830–1870"; and Lucier, "Commercial Interests and Scientific Disinterestedness: Consulting Geologists in Antebellum America," 245–267.

2. On the elder Silliman, see Brown, *Benjamin Silliman: A Life in the Young Republic* and Baatz, "'Squinting at Silliman.'"

3. No complete biography of the younger Silliman exists, but an overview of his life and career may be found in Kuslan, "Benjamin Silliman, Jr.: The Second Silliman," and Wright, "Benjamin Silliman, 1816–1885." On Silliman's early consulting work, see Lucier, "Commercial Interests," 251, 254, 258–259 n. 30.

4. Lucier, "Commercial Interests," 245–267.

5. Lucier, "Scientists and Swindlers," 349–386; Saltzman, "The Art of Distillation and the Dawn of the Hydrocarbon Society," 53–60.

6. Webb, "Mines in Northwestern Arizona," 250 n. 5, 251; Church, "Scott, Thomas Alexander," 500–501.

7. Webb, "Mines in Northwestern Arizona," 248–250, 257 n. 21, 262 n. 30; Thompson, "'Is There a Gold Field East of the Colorado?'" 345–363; Lingenfelter, *Steamboats on the Colorado River, 1852–1916*, 33–40. *Arizona Miner*, 11 May 1864.

8. White, "The Case of the Salted Sample," 153–154; White, *Scientists in Conflict*, 45–73; Webb, "The Chemist as Consultant in Gilded Age America," 10–11.

9. Webb, "Mines in Northwestern Arizona," 252; White, "Salted Sample," 154; Silliman, "Mining Districts of Arizona," 289–293.

10. Silliman, "Mining Districts of Arizona," 293–296, 300–308; Webb, "Mines in Northwestern Arizona," 257, 261–265. *Mining and Scientific Press*, 6 August, 3 December 1864; 9 November 1867; 19 September, 21 November 1868.

11. The definitive account of the southern California oil boom is White, *Scientists in Conflict*. For a discussion of Silliman's perspective on his work in southern California, see Webb, "Benjamin Silliman's Visit to California," 365–378.

12. *Mining and Scientific Press*, 28 January, 4 February 1865.

13. Webb, "Silliman's Visit," 366–368; Webb, "Chemist as Consultant," 12; White, *Scientists in Conflict*, 152–154; Lucier, "Scientists and Swindlers," 493–516.

14. *New York Times*, 13, 17 March 1865. Silliman, "Mining Districts of Arizona," 289, 294–297.

15. Webb, "Chemist as Consultant," 12; White, *Scientists in Conflict*, 109, 134–136.

16. Webb, "Chemist as Consultant," 12–13; White, *Scientists in Conflict*, 166–172. The standard work on the Emma Mine scandal remains Spence, *British Investments and the American Mining Frontier, 1860–1901*, 139–182.

17. *Engineering and Mining Journal*, 21 July 1877.

18. Zhu, "Historical Statistics of the New Mexico Mining Industry," 92–94; Twitchell, *Leading Facts of New Mexican History*, 2:482–491.

19. Swick, *Cultural Resource Survey of Turquoise Hill, Santa Fe County, New Mexico*, 21; Schroeder, "The Cerrillos Mining Area," 13–16. *Deed Book A*, Santa Fe County Mining Records, State Records Center and Archives, Santa Fe, New Mexico. Al Regensberg to author, 14 August 1997, author's files.

20. *The Grand Central Tunnel Mining Company of New Mexico*, 12–19.

21. Ibid., 21–23.

22. Ibid., 3–4, 11. LeBaron Bradford Prince to James Wadsworth, 5 November 1881, LeBaron Bradford Prince letters to James Wadsworth, Western Americana Collection, Beinecke Rare Book and Manuscript Library, Yale University, New Haven, Connecticut. On Prince, see Larson, *New Mexico's Quest for Statehood, 1846–1912*, 141–146.

23. Silliman, "Turquois of New Mexico," 67–71; Swick, *Cultural Resource Survey*, 19–22, 138.

24. Silliman, *The Rio Grande Gold Gravels*, 4–7, 13–22, 25–31.

25. Ibid., 7, 34.

26. Wright, "Benjamin Silliman," 132.

27. *Engineering and Mining Journal*, 17 September, 1 October, 26 November 1881; *Mining and Scientific Press*, 22 October 1881.

28. Silliman, "The Mineral Regions of Southern New Mexico," 424–426.

29. Jones, *Old Mining Camps of New Mexico, 1854–1904*, 28–34. *Mining and Scientific Press*, 1 October 1881. Homer E. Milford to author, 12 May 1998, author's files.

30. Silliman, "Mineral Regions of Southern New Mexico," 428–429, 434–436.

31. Sherman, *Ghost Towns and Mining Camps of New Mexico*, 132–133; Endlich, "The Mining Regions of Southern New Mexico," 151–152.

32. Silliman, "Mineral Regions of Southern New Mexico," 437–442. *Engineering and Mining Journal*, 17 June 1882.

33. *Engineering and Mining Journal*, 15 July 1882.

34. Ibid., 11 November 1882. Silliman, "Mineral Regions of Southern New Mexico," 442; Sherman, *Ghost Towns and Mining Camps of New Mexico*, 132–133; Bancroft, *History of Arizona and New Mexico, 1530–1888*, 754–755.

35. University of Arizona Library Special Collections, Arizona Biographical File, "George A. Treadwell." George A. Treadwell to Silliman, 16, 30 April, 21, 29 May, 25 June, 5 August, 27, 29 November, 18 December 1875, University of Arizona Library Special Collections, Silliman Correspondence, 1875–1884 [hereafter, Silliman Correspondence, UAL].

36. Bancroft, *History of Arizona and New Mexico*, 582–592. George A. Treadwell to Silliman, 19 May, 1 August 1880; 28 January, 26 February 1881, Silliman Correspondence, UAL. J. B. Treadwell to Silliman, 26 December 1880, ibid. Thomas F. Hopkins to Silliman, 10 May 1881, ibid.

37. Love, *Mining Camps and Ghost Towns*, 39–56, 87–93. On Frémont as territorial governor, see Wagoner, *Arizona Territory, 1863–1912: A Political History*, 164–193. George A. Treadwell to Silliman, 6 November 1880, Silliman Correspondence, UAL.

38. Love, *Mining Camps and Ghost Towns*, 88, 93–94; Silliman, "Mineralogical Notes," 198–199. F. F. Thomas to Silliman, 3 May 1881, Silliman Correspondence, UAL. George A. Treadwell to Silliman, 14 September, 23 December 1881, ibid. Diary, 6 April–1 May 1881, box 2, vol. 166, William Phipps Blake Papers, Arizona Historical Society, Tucson, Arizona.

39. George A. Treadwell to Silliman, 2, 6 April 1880; 8 November 1882, Silliman Correspondence, UAL.

40. Silliman, "Mineralogical Notes," 198–203. *Mining and Scientific Press*, 29 October 1881; *Engineering and Mining Journal*, 19 November 1881. George A. Treadwell to Silliman, 14 September 1881, Silliman Correspondence, UAL.

41. George A. Treadwell to Silliman, 19 September 1881; 11 March, 1 November, 29 December 1882; 25 March 1883, Silliman Correspondence, UAL. Dunning, *Rock to Riches*, 376.

42. Dunning, *Rock to Riches*, 77–79, 103–105. C. P. Head and Hugo Richards to Silliman, 7 November 1880, Silliman Correspondence, UAL. George A. Treadwell to Silli-

man, 6, 13, 26 November, 2 December 1880; 2, 4, 19 January, 5 February 1881, ibid. For a discussion of Phelps Dodge, see Schwantes, *Vision and Enterprise*.

43. Dunning, *Rock to Riches*, 78–79, 113–115; Peplow, *History of Arizona*, 2:224–228; Wagoner, *Arizona Territory*, 194, 202, 216–217, 222. George A. Treadwell to Silliman, 11 April, 2 May 1883, Silliman Correspondence, UAL.

44. George A. Treadwell to Silliman, 21, 27 May 1883; 5 January, 5 February 1884, Silliman Correspondence, UAL.

45. Wright, "Benjamin Silliman," 132–133.

46. Lucier, "Commercial Interests," 245–267. Lucier's comment concerning Silliman appears on p. 251.

Chapter 3. Tree Rings and Climatic Cycles

1. The standard biography of Douglass is Webb, *Tree Rings and Telescopes*.

2. Ibid., 1–53.

3. "Cycles: A Problem in Naming," 20 December 1934, folder 3, box 67; "Incidents," 4 April 1939, folder 1, box 93, University of Arizona Library Special Collections, A. E. Douglass Papers [hereafter, AED/UAL]. Webb, "Solar Physics and the Origins of Dendrochronology," 291–301.

4. Hufbauer, *Exploring the Sun*, 46–47, 52–79. Very, "The Variation of Solar Radiation," 255–272; Langley, "On a Possible Variation of the Solar Radiation," 305–321; Clough, "Synchronous Variations in Solar and Terrestrial Phenomena," 42–75; Abbot, "The Relation of the Sun-Spot Cycle to Meteorology," 178–181; Bigelow, "Studies on the Diurnal Periods," 292–295. "Notes from Reading, 1900?-1914," folder 1, box 102, AED/UAL.

5. Douglass, "Weather Cycles in the Growth of Big Trees," 225–237.

6. Webb, *Tree Rings and Telescopes*, 105–106. Douglass, *Climatic Cycles* (1919), 16–17, 27–29, 74.

7. Sketches and notes dated 4 April 1933, box 99, AED/UAL. Douglass, "Photographic Periodogram," 326–331; Douglass, "An Optical Periodograph," 173–186.

8. Douglass, *Climatic Cycles* (1919), 74–76, 81–84, 98–99.

9. Ibid., 74, 98–102.

10. F. E. Clements to Douglass, 1 December 1922 (telegram), box 80; Notes, Conference on the Origin and Nature of Cycles, 8–9 December 1922, folder 5, box 126, AED/UAL. "Report of a Conference on Cycles," *Geographical Review* 13 (Special Supplement, October 1923): 657–676.

11. Webb, *Tree Rings and Telescopes*, 114–119, 131–151.

12. Various notes, drafts, and correspondence, boxes 80, 109, AED/UAL.

13. A. E. Douglass, *Climatic Cycles* (1928), 118, 123, 125, 133, 137–138.

14. Ibid., 133.

15. John C. Merriam to Douglass, 6 November 1928, box 78, AED/UAL. "The Second Conference on Cycles," *Geographical Review* 19 (April 1929): 296–306.

16. John C. Merriam to Douglass, 6 February 1932; Douglass to Merriam, 15 Feb-

ruary 1932, box 78, AED/UAL. "Symposium on Climatic Cycles," National Academy of Sciences, *Proceedings* 19 (1933): 349–388.

17. Douglass, "Evidences of Cycles in Tree Ring Records," 350–360.

18. "Symposium on Climatic Cycles," 386, 388.

19. Ibid., 386–387.

20. Ibid., 387–388.

21. "The Carnegie Cooperation," folder 9, box 78; Douglass to W. S. Adams, 6 June, 24 September 1932, box 66; Various correspondence, 1932, folder 2, box 79; Douglass to H. A. Spoehr, 2 November 1932, box 80, AED/UAL.

22. Notes for Carnegie Conference, 13 October 1932, folder 5, box 66; Frederic E. Clements to Douglass, 31 October, 7 November 1932, box 80, AED/UAL.

23. "Sun-Spot Cycles," *Science* 76 (16 December 1932), Supplement, 8–9.

24. "Notes and Thoughts on Carnegie Institution," folder 9, box 78; Douglass to J. C. Merriam, 26 December 1932, box 78; Notes dated 26 August 1936, folder 1, box 93, AED/UAL.

25. Webb, *Tree Rings and Telescopes*, 155–158. J. C. Merriam to Douglass, 26 January, 1 February, 16 December 1935, box 78, AED/UAL.

26. Outline notes, dated Summer 1935, folder 1, box 111; Various drafts and correspondence, box 115, AED/UAL.

27. Edwin B. Wilson to Douglass, 19 January 1937, box 68, AED/UAL. Douglass, *Climatic Cycles* (1936), ix–x, 1, 5.

28. Douglass, *Climatic Cycles* (1936), 133–138.

29. Webb, *Tree Rings and Telescopes*, 158–161.

30. See, for example, H. T. Potts to Douglass, 15 July 1933, Perry R. Taylor to Douglass, 4 April 1934; F. C. Merriell to Douglass, 5 July 1934; Charles F. Sarle to Douglass, 8 August 1934, box 68, AED/UAL.

31. Douglass to Alfred Atkinson, 31 May 1940, box 86; Edmund Schulman to John T. Tucker, 2 June 1941, box 69, AED/UAL.

32. "Dramatic Incidents in Tree-Ring Studies," 18 April 1949, folder 1, box 93; Various notes, records, reports, and correspondence, box 92, AED/UAL.

33. See, for example, Fritts, *Reconstructing Large-scale Climatic Patterns from Tree-Ring Data.*

Part II. Introduction

1. Good overviews of the region's growth and significance include Meinig, *Southwest: Three Peoples in Geographical Change, 1600–1970*; Rothman, *On Rims & Ridges*; Luckingham, *The Urban Southwest*; Lowitt, *The New Deal and the West*; and Nash, *The American West Transformed.*

2. Rothman, *Rims & Ridges*, 39–55; Reid and Whittlesey, *Archaeology of Ancient Arizona*, 13, 176–178; Elliott, *Great Excavations*, 26, 54–55, 79–80, 103–106. Also see Hinsley and Wilcox, *The Southwest in the American Imagination*; Snead, *Ruins and Rivals*; and Fowler, *A Laboratory for Anthropology.*

3. On higher education in New Mexico, see Hughes, *Pueblo on the Mesa*; Kropp, *That All May Learn*; and Christiansen, *Of Earth and Sky*.

4. Arizona universities are discussed in Martin, *Lamp in the Desert* and Hopkins and Thomas, *The Arizona State University Story*.

5. Kropp, *That All May Learn*, 22–23, 382, 384; Rice, "Arizona Agricultural Experiment Station," 123–124, 127–128.

6. Rosenberg, *No Other Gods*, 154–174.

7. Visher, *Scientists Starred*, 3–5; Cattell, *American Men of Science*, v–vi.

8. See, for example, Lankford, *American Astronomy* and Rossiter, *Women Scientists in America*.

Chapter 4. Foundations and Institutions

1. Luckingham, *Urban Southwest*, 33–34, 46–47; Martin, *Lamp in the Desert*, 279; Hughes, *Pueblo on the Mesa*, 41, 45; Kropp, *That All May Learn*, 96, 116; Christiansen, *Of Earth and Sky*, 18–19, 41.

2. Rice, "Arizona Agricultural Experiment Station," 123–140; Kropp, *That All May Learn*, 99, 121; Rosenberg, *No Other Gods*, 154–174.

3. McGinnies, *Discovering the Desert*, 1–16; Bowers, *A Sense of Place*, 1–17, 31–33.

4. Rice, "Arizona Agricultural Experiment Station," 140; Harding, *Two Blades of Grass*, 118, 140; Lorbiecki, "The Land Makes the Man," 235–252.

5. Rothman, *On Rims & Ridges*, 56–124; Woodbury, *Sixty Years of Southwestern Archaeology*, 429. Valuable overviews of the history of Southwestern archaeology include Reid and Whittlesey, *Archaeology of Ancient Arizona*; Elliott, *Great Excavations*; and Fowler, *A Laboratory for Anthropology*.

6. Visher, *Scientists Starred*, 150, 270, 278.

7. Paterson, *Hot Empire of Chile*, 15–24. Examples of García's early work include "The effect of spring frosts on the peach crop" (1899); "Orchard notes" (1901); and "Spraying orchards for the codling moth" (1902).

8. García, "Chile Culture" (1908); García, "Improved variety no. 9 of native chile" (1921). Cotter, "Scientific Contribution of New Mexico to the Chile Pepper," 17–18; Paterson, *Hot Empire of Chile*, 16–19, 84–85.

9. García and Rigney, "Winter protection of the vinifera grape" (1916); García, "New Mexico beans" (1917); García and Fite, "Preliminary pecan experiments" (1925); Fite and García, "Pear variety test" (1928); Linney and García, "Climate in relation to crop adaptation in New Mexico" (1918); García and Fite, "Preliminary smudging experiments" (1922).

10. Luckingham, *Urban Southwest*, 48, 61–62; Martin, *Lamp in the Desert*, 150–153, 166–170, 279; Hughes, *Pueblo on the Mesa*, 45–46; Kropp, *That All May Learn*, 174–175, 195; Christiansen, *Of Earth and Sky*, 19, 41.

11. Bowers, *A Sense of Place*, 42–43, 56–59, 78–81; Alexander, "From Rule of Thumb to Scientific Range Management," 409–428.

12. Bowers, *A Sense of Place*, 127–128.

13. Various material in folder 1, box 131, AED/UAL.

14. Various material in boxes 131–132, ibid.

15. Cattell, "Origin and Distribution of Scientific Men," 513–516.

16. Shreve's career is surveyed in Bowers, *A Sense of Place*. On Slipher, see Hoyt, "Vesto Melvin Slipher," 411–449. The standard biography of Douglass is Webb, *Tree Rings and Telescopes*.

17. Martin, *Lamp in the Desert*, 279; Hughes, *Pueblo on the Mesa*, 46; Kropp, *That All May Learn*, 223, 243; Christiansen, *Of Earth and Sky*, 27–30, 41, 64–66; Sonnichsen, *Pass of the North*, 2:31, 35.

18. Evans and Mulholland, *Big and Bright*, 6–29, 33–90; Osterbrock, *Yerkes Observatory, 1892–1950*, 187–244.

19. Bowers, *A Sense of Place*, 91–98, 142–145; Webb, *Tree Rings and Telescopes*, 155–161.

20. "Tucson Gets Experiment Station," 467.

21. Luckingham, *Urban Southwest*, 67–69; Lowitt, *The New Deal and the West*, 17–21, 82–85, 122–137, 210; Martin, *Lamp in the Desert*, 165–166; Rothman, *On Rims & Ridges*, 177–206.

22. Kunitz, *Disease Change and the Role of Medicine*, 140–141, 146–178; Trennert, *White Man's Medicine*, 177–199; Schackel, *Social Housekeepers*, 61–85.

23. For a discussion of "relief archaeology," see Fagette, *Digging for Dollars*.

24. Reid and Whittlesey, *Archaeology of Ancient Arizona*, 208, 224–225; Elliott, *Great Excavations*, 72–73.

25. Fagette, *Digging for Dollars*, 134, 140–142, 152.

26. Additional demographic information concerning the Southwest's scientific community may be found in Webb, "Scientists in the American Southwest."

Chapter 5. Leading Women Scientists

1. Exceptions to this generalization are Babcock and Parezo, *Daughters of the Desert*; Parezo, *Hidden Scholars*; and Kofalk, *No Woman Tenderfoot*. The standard study of women scientists remains Rossiter, *Women Scientists in America*, but also see Lankford and Slavings, "Gender and Science." An earlier (1993) version of this chapter appeared as Webb, "Leading Women Scientists."

2. No southwestern women scientists appeared in the volumes published in 1906 and 1910. The 1944 volume, compiled during the war years, was omitted because of its unreliability concerning southwestern scientists, many of whom remained listed with their home institutions outside the region.

3. Rossiter, *Women Scientists in America*, xv–xviii, 15–65, 99–110, 204–212.

4. Ibid., 120–168.

5. Ibid., 52–71, 116–121, 137–138, 160–165, 200–204.

6. The new specialty of home economics had become visible in New Mexico somewhat earlier. See Jensen, "Canning Comes to New Mexico" and Lee and Brown, "Women at New Mexico State University."

7. Bowers, *A Sense of Place*, 21–23, 46–50, 66–67, 70, 82, 95. Shreve, *The Daily March of Transpiration in a Desert Perennial*. Another example of Shreve's research interests is Shreve, "Investigations on the Imbibition of Water by Gelatine."

8. Shreve, "Seasonal Changes in the Water Relations of Desert Plants," 266–292.

9. Rossiter, *Women Scientists in America*, 36–40, 146–159.

10. Ibid., 138, 238–241. Among southwestern women scientists in the 1920s and early 1930s, only chemist Lila Sands held a position in a traditionally male-dominated field.

11. Handschin, "Percentage of Women Teachers in State Colleges and Universities," 55–57.

12. Webb, "A Woman's Place is in the Lab," 57–61.

13. Mary Estill Caldwell biographical file, University of Arizona Library Special Collections, University of Arizona Faculty Files. Rossiter, *Women Scientists in America*, 187–188.

14. Recent discussions of New Deal programs concerning the health of Native Americans include Trennert, *White Man's Medicine*, 177–199 and Schackel, *Social Housekeepers*, 61–85.

15. Babcock and Parezo, *Daughters of the Desert*, 147. For background information on Hohokam archaeology, see Haury, *The Hohokam: Desert Farmers & Craftsmen*, 3–9. On the Hodges Site, see Hartmann, *The Hodges Ruin: A Hohokam Community in the Tucson Basin*.

16. Babcock and Parezo, *Daughters of the Desert*, 125–129.

17. The two M.D. degrees were from Yale and Rochester.

18. The literature on the Southwest's role as a tuberculosis haven includes many good studies. For the period before 1900, see Jones, *Health-Seekers in the Southwest*, 123–198. Valuable discussions of the situation in New Mexico include Spidle, *Doctors of Medicine in New Mexico*, 87–169; and Johnson, "John Weinzirl: A Personal Search for the Conquest of Tuberculosis," 141–155.

19. Bowers, *A Sense of Place*, 148. Edith Shreve did not abandon research in plant physiology, however, as shown by her June 1940 presentation at the Pacific Section meeting of the Botanical Society of America. Shreve, "The Relation of Transpiration to Evaporation from Artificial Surfaces," 707.

20. Babcock and Parezo, *Daughters of the Desert*, 125.

21. Ibid., 129, 135.

22. Ibid., 159. Tanner's lack of a Ph.D. slowed her promotion rate significantly.

23. Especially in higher education, administrators often applied antinepotism rules in a rigid fashion to avoid hiring women scientists. Rossiter, *Women Scientists in America*, 190–197, 216. This practice seems not to have been widely followed in the Southwest. For an explanation of the greater opportunities for women in western higher education, see Handschin, "Percentage of Women Teachers," 57.

Chapter 6. The Impact of World War II

1. USDA, *Agricultural Statistics 1936*, 131, *1940*, 210, *1946*, 181; USDA, *Yearbook of Agriculture 1937*, 755; *WPA Guide to 1930s Arizona*, 83–84.

2. USDA, *Yearbook of Agriculture 1937*, 760–804; "Yuma Agricultural Center," http://www.ag.arizona.edu/aes/yac/yac.htm; King, "Agricultural Investigations," 49–51; Martin, "Grapefruit Storage Studies," 529–534; Hilgeman and Smith, "Changes in Invert Sugar," 535–538; Jones, "Relation of Nitrogen Absorption," 1–4; Finch and McGeorge, "Studies of Grapefruit Fertilization," 62–67.

3. Roehm, "Vitamin B and G Content," 663–666; Jones, "Note on Ascorbic Acid," 103–104; Smith, "Comparative Vitamin C Values."

4. Nash, *American West Transformed*, 14. Also see Nash, *World War II and the West*.

5. Nash, *American West Transformed*, 222–223; Luckingham, *Urban Southwest*, 75–79; Sheridan, *Arizona*, 272–273.

6. Luckingham, *Urban Southwest*, 79–94; Sheridan, *Arizona*, 278–282; Simmons, *Albuquerque*, 369–370.

7. Nash, *American West Transformed*, 153–177; Simmons, *Albuquerque*, 367.

8. Hoddeson, *Critical Assembly*, 24–31, 40–66. Other valuable studies of Los Alamos include Kunetka, *City of Fire*; Rhodes, *The Making of the Atomic Bomb*; *Project Y: The Los Alamos Story*; and Szasz, *The Day the Sun Rose Twice*.

9. Kunetka, *City of Fire*, 4; *Project Y*, xvii, 484; Welsh, *U.S. Army Corps of Engineers, Albuquerque District, 1935–1985*, 83, 86–92.

10. For details of the atomic bomb project, see Hoddeson, *Critical Assembly*; Rhodes, *Making of the Atomic Bomb*; and Szasz, *The Day the Sun Rose Twice*.

11. Kunetka, *City of Fire*, 190–208; Hoddeson, *Critical Assembly*, 398–402; *Project Y*, 265–277, 295.

12. *Project Y*, 353–368, 380–430; Rothman, *On Rims & Ridges*, 233–257.

13. Furman, *Sandia National Laboratories*, 119–141, 147–205, 214–225, 231–258, 307–362, 453.

14. DeVorkin, *Science With A Vengeance*, 45–107, 109–133; Ordway and Sharpe, *Rocket Team*, 277–278, 344–346.

15. Ordway and Sharpe, *Rocket Team*, 308–317, 346–349.

16. Ibid., 349–355; DeVorkin, *Science With A Vengeance*, 63–67, 135–149.

17. DeVorkin, *Science With A Vengeance*, 111–113, 140.

18. Ibid., 168–182, 197–220, 247–252, 258–262; Ordway and Sharpe, *Rocket Team*, 356, 363, 366; Holliday, "Seeing the Earth from 80 Miles Up," 511–528.

19. Nash, *World War II and the West*, 164–165.

20. Woodbury, *Sixty Years of Southwestern Archaeology*, 147–148; Reid and Whittlesey, *Archaeology of Ancient Arizona*, 46–47, 209–228.

21. Reid and Whittlesey, *Archaeology of Ancient Arizona*, 17–18, 138–147. For a discussion of the field school, see Haury, *Point of Pines, Arizona*.

22. Woodbury, *Sixty Years of Southwestern Archaeology*, 149–166, 187–212; Stocking,

"The Santa Fe Style in American Anthropology," 15; Reid and Whittlesey, *Archaeology of Ancient Arizona*, 19.

23. Evans and Mulholland, *Big and Bright*, 90–92, 98–105, 178 n. 98; Osterbrock, *Yerkes Observatory, 1892–1950*, 258–259; Doel, *Solar System Astronomy*, 46–48.

24. Evans and Mulholland, *Big and Bright*, 106–110, 117–120; Osterbrock, *Yerkes Observatory, 1892–1950*, 279–301; Doel, *Solar System Astronomy*, 45–57.

25. Doel, *Solar System Astronomy*, 68–76.

26. Ibid., 57–68.

27. U.S. Office of Education, "Statistics of Higher Education, 1939–40 and 1941–42," *Biennial Surveys of Education in the United States, 1938–40, 1940–42* (Washington, D.C.: Government Printing Office, 1947), 2:101, 130–131, 146; Hansen, *The World Almanac and Book of Facts* (1950), 545–562.

28. Christiansen, *College on the Rio Grande*, 48–55, 143.

Part III. Introduction

1. Masse, "Prehistoric Irrigation systems in the Salt River Valley, Arizona," 408–415; Neilson, "High-Resolution Climatic Analysis and Southwest Biogeography," 27–34; Schlesinger, "Biological Feedbacks in Global Desertification," 1043–1048. On the Langmuir Laboratory, see Chew, *Storms Above the Desert*.

2. Webb, "Dendrochronology," 164–166.

3. Lantz, "Carotene and Ascorbic Acid Contents of Peppers"; Lantz, "Some Factors Affecting the Ascorbic Acid Content of Chile"; Lantz, "Effects of Canning and Drying." Edith M. Lantz Biographical Form, Rio Grande Historical Collections/Hobson-Huntsinger University Archives, New Mexico State University, Las Cruces, New Mexico.

4. Cotter, "Scientific Contribution of New Mexico to the Chile Pepper," 18–21; Paterson, *Hot Empire of Chile*, 17–22, 27–34; Bosland and Votava, "Chile Cultivars," 2–4; Proulx, "Some Like Them Hot," 52–54.

5. Cotter, "Scientific Contribution," 17–18; Paterson, *Hot Empire*, 84–88.

6. Szasz, "The Cultures of Modern New Mexico," 167–168; Chapin, "Garrett and Pressurized Flight," 335, 337, 340–342.

7. Szasz, "Cultures of Modern New Mexico," 169; Fisher, "Aviation Medicine on the Threshold of Space," 258–259, 262, 271; Weaver, "Countdown for Space," 725–727; Gilruth, "The Making of an Astronaut," 124–127, 144.

8. Webb, "Kitt Peak National Observatory," 296–297. Also see Edmondson, *AURA and Its US National Observatories*.

9. Tatarewicz, *Space Technology and Planetary Astronomy*, 26–46, 78–80, 148–151; Whitaker, *The University of Arizona's Lunar and Planetary Laboratory*, 33–36.

10. Webb, "University of Arizona, Astronomy at," 543–545; Webb, "Multiple Mirror Telescopes," 338–340.

11. Finley, "Radio Astronomy in New Mexico: The VLA and VLBA," 21–26; "New Mexico Funds New NRAO Building," 8–9.

12. Waldrop, "New Art of Telescope Making," 1495–1497; Martin, New Ground-Based Optical Telescopes," 22–30; Anderson, "To the Edge of the Universe," 6–20.

13. Marshall, "Supercollider Sweepstakes," 1288; Flam, "Is There Life After the SSC?" 644–647; Goodwin, "Amazing Race," 69–74; Goodwin, "After Agonizing Death in the Family, Particle Physics Faces Grim Future," 87–91.

14. Szasz, "Cultures of Modern New Mexico," 169–171; *Albuquerque Journal*, 23, 26, 31 August, 21 September 1996.

15. "Mt. Graham Site Faces Opposition," 1, 4; Waldrop, "The Long, Sad Saga of Mount Graham," 1479–1481; Travis, "Scopes and Squirrels Return to Court," 1356; Mervis, "Red Squirrels 2, Astronomers 0," 630; "Green Light for Mount Graham Telescope?" 637. Also see Istock and Hoffman, *Storm over a Mountain Island*.

16. Szasz, "Cultures of Modern New Mexico," 171–172; Anderson, "Weapons Labs in a New World," 168–171; Goodwin, "To Replace AT&T at Sandia, DOE Picks Martin Marietta," 53–54.

17. Waldrop, "Troubled Times Ahead for Telescope-Makers," 28–31; Kaiser, "Plan Would Shut Kitt Peak Facilities," 641; Waldrop, "Radio Astronomy's Crumbling Showpiece," 268–269.

Chapter 7. Astronomy in Southern Arizona

1. Martin, *Lamp in the Desert*, 20–46; Douglass, "Historical Address," 54.

2. Tucson *Citizen*, 14 February, 10 June 1907; Douglass, "Historical Address," 54; Douglass, "Drawings of Comet *a*, 1910," 162–163. W. H. Pickering to Douglass, 11 March 1900, Douglass to E. C. Pickering, 13 May 1908; Willard P. Gerrish to Douglass, 9 March 1909, box 44, AED/UAL.

3. George E. Hale to Douglass, 17 July 1914; E. B. Frost to Douglass, 20 July 1914; Douglass to J. A. Brashear, 15 June 1914; to John T. Hughes, 15 June 1914; to M. P. Freeman, 6 July 1914; to George W. P. Hunt, 3 August 1914; "Plans 1914," box 44, AED/UAL. Tucson *Citizen*, 29 July, 30 September, 28, 31 October 1914; Arizona *Star*, 10 October 1914.

4. Arizona *Star*, 19 October 1916, 12 August 1917; A. E. Douglass, "Steward Observatory," *Annual Report of the University of Arizona* (1939): 7. Douglass to W. J. Hussey, 24 July 1916; to George E. Hale, 26 July 1916; to Warner and Swasey, 19 August 1916; "Plans for Immediate Operation of the Steward Observatory"; Report to President, 10 November 1916, box 44, AED/UAL.

5. Webb, *Tree Rings and Telescopes*, 63–74. A more detailed study of the establishment of the Steward facility is Webb, "The Indefatigable Astronomer."

6. Douglass, "The University of Arizona Eclipse Expedition," 170–184. Steward Observatory Annual Report (1922–1923, 1923–1924), box 47, AED/UAL. The ASP was one of several groups meeting with the American Association for the Advancement of Science in the largest scientific conference held in the West to that date.

7. Steward Observatory Annual Report (1923–1924, 1924–1925), box 47; various Martian notes, folder 5, box 53, AED/UAL.

8. Douglass to F. C. Lockwood, 4 June 1924; Steward Observatory Annual Report (1924–1925); Report of Astronomy Department (1925–1926), box 47, AED/UAL.

9. Carpenter, "U Cephei: An Anomalous Spectrographic Result," 205–220.

10. Steward Observatory Annual Report (1928–1929), box 47, AED/UAL. International Astronomical Union, *Transactions* 2 (1925): 132–141; 3 (1928): 193–195.

11. Carpenter, "The Distribution of Color in Two Extra-Galactic Nebulae," 294–295; Carpenter, "A Cluster of Extra-Galactic Nebulae in Cancer," 247–254; Hubble and Humason, "The Velocity-Distance Relation Among Extra-Galactic Nebulae," 63 n. 1.

12. Carpenter, "Note on the Absorption of Radiation in Space," 25; *Arizona Daily Star*, 9 September 1931; *New York Times*, 10, 13 September 1931.

13. Steward Observatory Report (29 November 1928), box 47; Douglass to H. L. Shantz, 9 May 1930; to H. A. Spoehr, 20 April 1936; to John Collier, 20 October 1938; G. A. Moskey to Douglass, 3 November 1938; Theodore B. Hall to Douglass, 16 November 1938, box 48, AED/UAL. *Arizona Daily Star*, 4 May 1932.

14. *University of Arizona Record* (1936): 21; (1938): 18, 162–163. Astronomy Department Report (1937–1938), box 47, AED/UAL. On Douglass and dendrochronology, see Webb, *Tree Rings and Telescopes*, 101–190.

15. Steward Observatory Report (1940), University of Arizona Library Special Collections, Reports to President [hereafter Reports/UAL]. Roach and Stoddard, "A Photoelectric Light-Curve of Eros," 305–312; Roach, "A Composite Light-Curve of Eros," 310–313.

16. Steward Observatory Report (1940), Reports/UAL. Abt, "Award of the Bruce Gold Medal to Professor Willem J. Luyten," 247–251; Luyten, "The Search for White Dwarfs," 86–89; Carpenter, Deutsch, and Luyten, "Determination of Color Classes for 204 Stars of Large Proper Motion," 587–591.

17. Various notes, reports, and memoranda, boxes 47–48, AED/UAL. Webb, *Tree Rings and Telescopes*, 172–173.

18. Steward Observatory Report (1947, 1948, 1949, 1951), Reports/UAL. Wood, "The Eclipsing Variable RX Herculis," 465–474.

19. Luyten, "Search for White Dwarfs," 86, 88; Carpenter, "Determination of Color Classes," 591; Luyten and Carpenter, "First Report on a Systematic Survey for Faint Blue Stars," 429–431. Steward Observatory Report (1945–1957), Reports/UAL.

20. Steward Observatory Report (1951, 1954, 1955, 1956), Reports/UAL. Fitch, "The Cluster-Type Variable VZ Cancri," 690–710.

21. Steward Observatory Report (1952, 1954), Reports/UAL. Edmondson, *AURA and Its US National Observatories*, 6–8, 14–25.

22. Edmondson, *AURA and Its US National Observatories*, 26–45; *Tucson Daily Citizen*, 7 January 1954; *Arizona Daily Star*, 12 November 1955, 14 December 1957.

23. Steward Observatory Report (1957), Reports/UAL. *Arizona Daily Star*, 28 April, 8 September 1956.

24. Steward Observatory Report (1958), Reports/UAL. Fitch, "Photoelectric Observations of the Cluster-Type Variables EH Librae and DY Herculis," 108–111; Fitch, "The

Light-Variation of CC Andromedae," 701–715; Carpenter, "Spectra of Binary Galaxies," 386–388.

25. Edmondson, *AURA and Its US National Observatories*, 46–58, 85–93.

26. Steward Observatory Report (1958, 1959), Reports/UAL. The University of Arizona became a member of AURA in 1972. Edmondson, *AURA and Its US National Observatories*, 248.

27. Steward Observatory Report (1959, 1960), Reports/UAL. *Arizona Daily Star*, 23, 28 January 1959; 5, 16 March, 26 May 1960; Edmondson, *AURA and Its US National Observatories*, 94–106.

28. Steward Observatory Report (1960, 1961, 1962); Department of Astronomy Report (1960, 1961, 1963), Reports/UAL. *Arizona Daily Star*, 10 April 1960; 13 June, 22 August, 27 October 1961; 14, 25 February, 1, 6, 24 August, 5 September, 14 November 1962.

29. Steward Observatory Report (1960); Department of Astronomy Report (1960, 1961), Reports/UAL. *Arizona Daily Star*, 11 January, 2 September, 3 November 1961; *Tucson Daily Citizen*, 26 January 1962.

30. Webb, *Tree Rings and Telescopes*, 188–189; *Arizona Daily Star*, 26 January, 13 February 1963.

Chapter 8. Darwin in the Desert

1. Valuable discussions of the American debates concerning evolution include Larson, *Trial and Error* and *Summer for the Gods*; Numbers, *The Creationists*; Toumey, *God's Own Scientists*; and Webb, *Evolution Controversy in America*.

2. *Arizona Daily Star*, 12 January 1924; *Tucson Citizen*, 10 January 1924. For a more detailed account of Tucson's evolution controversy in the 1920s, see Webb, "Tucson's Evolution Debate, 1924–1927," 1–12.

3. *Tucson Citizen*, 9 January 1924; *Arizona Daily Star*, 9 January 1924.

4. "Beal, Richard S.," University of Arizona Library Special Collections, Arizona Biographical Files. *Tucson Citizen*, 11, 18 January 1924. Also see ibid., 2, 9 March, 19 May 1924.

5. *Arizona Daily Star*, 30 January 1925; *Arizona Republican*, 16–19 May 1925.

6. *Arizona Daily Star*, 17, 19 May 1925.

7. *Arizona Republican*, 26 May, 9, 10, 14, 22, 26, 27, 29 June, 7–9 July 1925; *Arizona Daily Star*, 7, 27 June 1925.

8. *Arizona Republican*, 7–9 July 1925; *Arizona Daily Star*, 4–9 July 1925.

9. *Arizona Republican*, 10–22 July 1925.

10. *Arizona Daily Star*, 10–12, 14–19, 21–23, 28 July 1925; *Arizona Republican*, 28 July–1 August 1925.

11. Webb, *Evolution Controversy in America*, 93–108.

12. "Beal, Richard S.," Arizona Biographical Files. *Arizona Daily Star*, 11 October 1927; Numbers, *The Creationists*, 57–60, 367 n. 15.

13. Numbers, *The Creationists*, 58–60, 366 n. 7, 8; *Arizona Daily Star*, 11, 13 October 1927.

14. *Arizona Daily Star*, 13, 16, 17 October 1927; *Arizona Republican*, 13 October 1927.

15. *Arizona Daily Star*, 19, 21–23 October 1927; *Arizona Republican*, 21 October 1927; Webb, "Tucson's Evolution Debate," 10–11.

16. *Arizona Republican*, 31 October, 2, 3, 5, 6 November 1927.

17. Webb, *Evolution Controversy in America*, 109–114, 128–134.

18. Lisonbee, "Thwarting the Anti-Evolution Movement in Arizona," 35; *Arizona Republic*, 28 January 1962. For the Mormon perspective on evolution, see Jeffrey, "Seers, Savants and Evolution," 41–75.

19. Grobman, *The Changing Classroom*, 205; *Phoenix Gazette*, 18 December 1963; *Arizona Republic*, 7 November 1963.

20. *New York Times*, 31 May 1964; *Arizona Daily Star*, 8 November 1963; *Phoenix Gazette*, 18, 21 December 1963; *Arizona Republic*, 18, 19 December 1963.

21. *Arizona Republic*, 19 December 1963, 23 May 1964; *New York Times*, 31 May 1964; Lisonbee, "Thwarting the Anti-Evolution Movement," 35.

22. *New York Times*, 31 May 1964; *Arizona Republic*, 23 May, 13, 28 June 1964; Lisonbee, "Thwarting the Anti-Evolution Movement," 36.

23. *Arizona Republic*, 23 May 1964; *New York Times*, 31 May 1964; *Arizona Daily Star*, 12 July 1964.

24. *Arizona Daily Star*, 12 July 1964; Wilhelm, "Chronology and Analysis of Regulatory Actions," 407, 410.

25. Webb, *Evolution Controversy in America*, 135–200, 206–235.

26. *Arizona Daily Reporter*, 8 February 1982; "News Briefs," *Creation/Evolution* 3 (Spring 1982): 43–44.

27. *Tucson Citizen*, 7 January 1982; *Arizona Daily Star*, 12, 13 February 1982; Edwords, "Creation-Evolution Debates: Who's Winning Them Now?" 37–38.

28. *Tucson Citizen*, 23 April 1983; "News Briefs," *Creation/Evolution* 4 (Spring 1983): 35.

29. *Arizona Republic*, 5 February 1987; Webb, *Evolution Controversy in America*, 236–237, 246–252.

30. Wuethrich, "Scientists Strike Back against Creationism," 659; Scott, "Not (Just) in Kansas Anymore," 813, 815; "Darwin's Brush With Racism," 1295; "Oklahoma Lawmakers," 431.

31. Ferber, "GM Crops," 1662–1666; Deloria, *Red Earth, White Lies*, 10–11; Morell, "Kennewick Man's Trials," 190, 192; McDonald, "Researchers Battle for Access," A18–A19, A22.

32. Koshland, "Dealing in Hot Property," 1585; "DOE won't run tests," 7; Wheelwright, "For our nuclear wastes," 42–43, 46–47; *New York Times*, 14 May 1998, 26 March 1999.

33. Webb, *Evolution Controversy in America*, 253–263; "Letters to the Editor," *Creation/Evolution* 27 (Summer 1990): 40–41; Matsumura, "Does Arizona Charter School

Teach Creationism?" 6; Matsumura, "Arizona: Another Sunbelt State Axes Evolution," 4.

Chapter 9. The Southwest and Interplanetary Exploration

1. Valuable surveys of the American space program include Burrows, *This New Ocean*; McDougall, . . . *the Heavens and the Earth*; and Lewis, *Appointment on the Moon*.

2. Hall, *Lunar Impact*, 3–10, 14–24, 53–54.

3. On Kuiper, see Cruikshank, "Gerard Peter Kuiper," 259–295.

4. Whitaker, *The University of Arizona's Lunar and Planetary Laboratory*, 14–19, 23.

5. Tatarewicz, *Space Technology and Planetary Astronomy*, xi–xiii, 13–56, 72–73, 89–91, 148–151; Whitaker, *Lunar and Planetary Laboratory*, 23–28.

6. Wilhelms, *To a Rocky Moon*, 20–22, 40–52, 57–59.

7. Hall, *Lunar Impact*, 63–80; Koppes, *JPL and the American Space Program*, 109–112, 117–122; Wilhelms, *To a Rocky Moon*, 94–110.

8. Whitaker, *Lunar and Planetary Laboratory*, 30; Tatarewicz, *Space Technology*, 56. Lunar and Planetary Laboratory, Annual Report, 1961–1962, University of Arizona Library Special Collections [hereafter, LPL/UAL].

9. Hall, *Lunar Impact*, 111–155, 163–182; Koppes, *JPL*, 122–133.

10. Hall, *Lunar Impact*, 185–198, 223–239; Wilhelms, *To a Rocky Moon*, 95; Hall, *Project Ranger: A Chronology*, 337.

11. Hall, *Lunar Impact*, 199–222; Koppes, *JPL*, 149–151.

12. Hall, *Lunar Impact*, 240–255; Koppes, *JPL*, 152–154.

13. Hall, *Lunar Impact*, 256–270; Koppes, *JPL*, 161–162; Wilhelms, *To a Rocky Moon*, 96–97.

14. Whitaker, *Lunar and Planetary Laboratory*, 47–48; Hall, *Ranger Chronology*, 463.

15. Hall, *Ranger Chronology*, 469–470, 476–477, 485, 487–489; Koppes, *JPL*, 163–164.

16. Hall, *Lunar Impact*, 280–302; Whitaker, *Lunar and Planetary Laboratory*, 48–49; Wilhelms, *To a Rocky Moon*, 105. Annual Reports, 1964–1965, 1965–1966, LPL/UAL.

17. Wilhelms, *To a Rocky Moon*, 139–149, 232–239, 309–310, 316–317; Whitaker, *Lunar and Planetary Laboratory*, 50.

18. Annual Reports, 1969–1970, 1970–1971, 1971–1972, LPL/UAL. Morrison and Samz, *Voyage to Jupiter*, 11, 23–25; Smith, "Voyage of the Century," 52, 56.

19. Morrison and Samz, *Voyage to Jupiter*, 35–37, 39; Smith, "Voyage of the Century," 52–53.

20. Annual Reports, 1973–1974, 1974–1975, LPL/UAL. *Tucson Daily Citizen*, 3 October 1974.

21. Burgess, *Far Encounter*, 17–20; Morrison and Samz, *Voyage to Jupiter*, 63–115.

22. Morrison and Samz, *Voyage to Jupiter*, 65–85; Terrile and Beebe, "Summary of Historical Data," 948–951.

23. Morrison and Samz, *Voyage to Jupiter*, 110. For a good overview of knowledge of Jupiter, see Beebe, *Jupiter: The Giant Planet*.

24. Morrison and Samz, *Voyage to Jupiter*, 139–167; Beebe, *Jupiter*, 93–149; Smith, "The Jupiter System Through the Eyes of Voyager 1," 956–969.

25. Morrison and Samz, *Voyage to Jupiter*, 84–85, 110–111, 136–137; Beebe, *Jupiter*, 150–167.

26. Burgess, *Far Encounter*, 17–18, 37.

27. Ibid., 30–31; Morrison, *Voyages to Saturn*, 50–93; Gore, "Voyager 1 at Saturn: Riddles of the Rings," 14, 17, 20–21, 24, 26–28.

28. Morrison, *Voyages to Saturn*, 86, 93; Burgess, *Far Encounter*, 27–29; Gore, "Voyager 1 at Saturn," 4–5, 10–11, 28–29.

29. Smith, "A New Look at the Saturn System: The Voyager 2 Images," 504–537; Morrison, *Voyages to Saturn*, 96, 99, 117–132.

30. Morrison, *Voyages to Saturn*, 100, 104, 109, 112–113, 115–116, 129–131, 157–158; Stone and Miner, "Voyager 2 Encounter with the Saturnian System," 499–504; Burgess, *Far Encounter*, 29.

31. Washburn, "Goodbye, Voyager," 45–46.

32. Gore, "Uranus: Voyager Visits a Dark Planet," 179–184, 190–191; Stone and Miner, "The Voyager 2 Encounter with the Uranian System," 39–43.

33. Burgess, *Far Encounter*, 33–35, 37; "Voyager," University of Arizona *Report on Research* 4 (Spring/Summer 1987): 7.

34. Smith, "Voyager 2 in the Uranian System: Imaging Science Results," 43–64; Gore, "Uranus," 180, 186–188, 194.

35. Burgess, *Far Encounter*, 10–11, 41, 48, 53–54; Smith, "Voyager 2 at Neptune: Imaging Science Results," 1422–1449.

36. Burgess, *Far Encounter*, 59–72; Smith, "Voyager 2 at Neptune," 1422–1431; Gore, "Neptune: Voyager's Last Picture Show," 35–41.

37. Smith, "Voyager 2 at Neptune," 1437–1448; Gore, "Neptune," 38–40, 45, 47; Burgess, *Far Encounter*, 137; Broadfoot, "Ultraviolet Spectrometer Observations of Neptune and Triton," 1459–1466.

38. Smith, "Voyager 2 at Neptune," 1431–1437; Gore, "Neptune," 37–38, 40; Porco, "An Explanation for Neptune's Ring Arcs," 995–1001.

39. Washburn, "Goodbye, Voyager," 48.

40. Ibid., 46; Beebe and Chanover, "Atmospheres of the Giant Planets," 94–95, 98–101; Porco, "Neptune's Ring Arcs," 995, 999–1000.

41. Lawler, "Small Missions Lift Planetary Science," 1596–1598; Matijevic, "Autonomous Navigation and the *Sojourner* Microrover," 454; Chaikin, "Hard Landings," 53. Information concerning the Pathfinder mission is readily available at the NASA/JPL Mars Exploration Program website: mpfwww.jpl.nasa.gov/.

42. Chaikin, "Hard Landings," 53–55; Kerr, "Gambling On a Martian Landing Site," 347–348. For an intriguing discussion of numerous aspects of the Mars Pathfinder mission, see Shirley, *Managing Martians*.

43. Newcott, "Return to Mars," 17, 20. Lori Stiles, "IMP to send view from Mars

on Independence Day," University of Arizona news release, 23 June 1997; Stiles, "UA Team Heads to Florida for Mars Pathfinder Launch," University of Arizona news release, 22 November 1996.

44. Stiles, "UA Scientists Show Mars to the World," 12–13. Stiles, "IMP to send view" (23 June 1997).

45. Kerr, "Ancient Life on Mars?" 864–866; Lawler, "Finding Puts Mars Exploration on Front Burner," 865; Scott, "Was There Life in the Martian Meteorite?" 8–9. A good overview of this topic is Goldsmith, *The Hunt for Life on Mars*.

46. Newcott, "Return to Mars," 17; Stiles, "UA Scientists Show Mars to the World," 12–13. Stiles, "IMP to send view" (23 June 1997). Details of IMP may be found at the Lunar and Planetary Laboratory website: www.lpl.arizona.edu.

47. Lori Stiles, "IMP team leader Peter Smith to visit President Clinton," University of Arizona news release, 16 July 1997. Newcott, "Return to Mars," 21.

48. See, for example, Schofield, "The Mars Pathfinder Atmospheric Structure Investigation/Meteorology (ASI/MET) Experiment," 1752–1758; and Hviid, "Magnetic Properties Experiments on the Mars Pathfinder Lander: Preliminary Results," 1768–1770.

49. Smith, "Results from the Mars Pathfinder Camera," 1758–1765. Lori Stiles, "IMP begins astronomy from Mars," University of Arizona news release, 9 July 1997.

50. Matijevic, "Autonomous Navigation," 454; Smith, "Mars Pathfinder Camera," 1758–1762; Kerr, "Pathfinder Tells a Geologic Tale With One Starring Role," 175.

51. Kerr, "Pathfinder Tells a Geologic Tale," 175; Rover Team, "Characterization of the Martian Surface Deposits by the Mars Pathfinder Rover, Sojourner," 1765–1768; Kerr, "Rocky Mix Suggests Wet Early Mars," 380; Kerr, "Possible Glimpse of Earth-like Geology In Mars Rock," 638–639; Golombek, "A Message from Warmer Times," 1470–1471.

52. Albee, "Mars Global Surveyor Mission: Overview and Status," 1671–1672; Malin, "Early Views of the Martian Surface from the Mars Orbiter Camera of Mars Global Surveyor," 1681–1685; Christensen, "Results from the Mars Global Surveyor Thermal Emission Spectrometer," 1692–1698.

53. Binder, "Lunar Prospector: An Overview," 1475–1476; Irion, "Lunar Prospector Probes Moon's Core Mysteries," 1423–1424; Feldman, "Fluxes of Fast and Epithermal Neutrons from Lunar Prospector," 1496–1500; Kerr, "Cheapest Mission Finds Moon's Frozen Water," 1628–1629.

Bibliography

Archival and Manuscript Sources

Arizona Historical Society, Tucson
> William Phipps Blake Papers

New Mexico State Records Center and Archives, Santa Fe
> Santa Fe County Mining Records

Rio Grande Historical Collections/Hobson-Huntsinger University Archives, New
> Mexico State University Library, Las Cruces
>> Faculty/Staff Biographical Files

University of Arizona Library Special Collections, Tucson
> Arizona Biographical Files
> A. E. Douglass Papers
> Reports to President
> Silliman Correspondence
> University of Arizona Faculty Files

Western Americana Collection, Beinecke Rare Book and Manuscript Library,
> Yale University Library, New Haven, Connecticut
>> LeBaron Bradford Prince, Letters to James Wadsworth

Periodicals and Serials

American Institute of Mining Engineers, *Transactions*, 1881–1882
American Journal of Science, 1866–1886
American Museum of Natural History, *Bulletin*, 1886–1914
American Naturalist, 1867–1883
Arizona Daily Star (Tucson), 1906–1995
Arizona Miner (Prescott), 1864
Arizona Republic (Phoenix), 1925–1987
Astronomical Society of the Pacific, *Publications*, 1920–1960
Astrophysical Journal, 1898–1960
Auk, 1890–1925
Engineering and Mining Journal, 1877–1882
Geographical Review, 1920–1930
Mining and Scientific Press, 1864–1881
Monthly Weather Review, 1900–1910
New York Times, 1865–1999
Science, 1900–2001
Tucson Citizen, 1906–1995
U.S. Department of Agriculture, *Agricultural Statistics*, 1930–1950
U.S. Department of Agriculture, *Yearbook of Agriculture*, 1910–1950

Books, Dissertations, and Pamphlets

Ackland, Len. *Making a Real Killing: Rocky Flats and the Nuclear West*. Albuquerque: University of New Mexico Press, 1999.
Allen, John Logan, ed. *North American Exploration*. Lincoln: University of Nebraska Press, 1997.
Archaeology and History of Santa Fe Country. Santa Fe: New Mexico Geological Society, 1997.
"Arizona and its Heritage." University of Arizona *Bulletin*, 7 (1936).
Aveni, Anthony F., ed. *Archaeoastronomy in the New World*. Cambridge: Cambridge University Press, 1982.
——, ed. *Archaeoastronomy in Pre-Columbian America*. Austin: University of Texas Press, 1975.
——, ed. *Native American Astronomy*. Austin: University of Texas Press, 1977.
Babcock, Barbara A., and Nancy J. Parezo. *Daughters of the Desert: Women Anthropologists and the Native American Southwest, 1880–1980*. Albuquerque: University of New Mexico Press, 1988.
Bancroft, Hubert Howe. *History of Arizona and New Mexico, 1530–1888*. Albuquerque: Horn & Wallace Publishers, 1962 (orig. pub. 1889).
Bartlett, Richard A. *Great Surveys of the American West*. Norman: University of Oklahoma Press, 1962.

Beebe, Reta. *Jupiter: The Giant Planet*. 2d ed. Washington, D.C.: Smithsonian Institution Press, 1997.

Bolton, Herbert Eugene. *Coronado: Knight of Pueblo and Plains*. Albuquerque: University of New Mexico Press, 1990 (orig. pub. 1949).

———. *Rim of Christendom: A Biography of Eusebio Francisco Kino, Pacific Coast Pioneer*. Tucson: University of Arizona Press, 1984 (orig. pub. 1936).

Bonta, Marcia Myers. *Women in the Field: America's Pioneering Women Naturalists*. College Station: Texas A & M University Press, 1991.

Bosland, Paul W., and Eric Votava. "The Chile Cultivars of New Mexico State University Released from 1913 to 1993." New Mexico State University, Agricultural Experiment Station, *Research Report 719* (1997).

Bowers, Janice Emily. *A Sense of Place: The Life and Work of Forrest Shreve*. Tucson: University of Arizona Press, 1988.

Brown, Michael Chandos. *Benjamin Silliman: A Life in the Young Republic*. Princeton: Princeton University Press, 1989.

Bruce, Robert V. *The Launching of Modern American Science, 1846–1876*. New York: Alfred A. Knopf, 1987.

Burgess, Eric. *Far Encounter: The Neptune System*. New York: Columbia University Press, 1991.

Burrows, William E. *This New Ocean: The Story of the First Space Age*. New York: Random House, 1998.

Cattell, J. McKeen, ed. *American Men of Science: A Biographical Directory*. Lancaster, Pennsylvania: The Science Press, 1906.

Chew, Joe. *Storms Above the Desert: Atmospheric Research in New Mexico, 1935–1985*. Albuquerque: University of New Mexico Press, 1987.

Christiansen, Paige W. *College on the Rio Grande: The Story of a Small School*. Socorro: New Mexico Institute of Mining and Technology, 1989.

———. *Of Earth and Sky: A History of New Mexico Institute of Mining and Technology, 1889–1964*. Socorro: New Mexico Institute of Mining and Technology, 1964.

Colley, Charles C. *The Century of Robert H. Forbes*. Tucson: Arizona Historical Society, 1977.

Colton, Harold S. *Hopi Katchina Dolls*. Rev. ed. Albuquerque: University of New Mexico Press, 1977 (orig. pub. 1959).

Coues, Elliott. *Birds of the Colorado Valley*. Washington, D.C.: U.S. Government Printing Office, 1878.

Cutright, Paul Russell, and Michael J. Brodhead. *Elliott Coues: Naturalist and Frontier Historian*. Urbana: University of Illinois Press, 1981.

Deloria, Vine, Jr. *Red Earth, White Lies: Native Americans and the Myth of Scientific Fact*. New York: Scribner, 1995.

Dethloff, Henry C., and Irvin M. May Jr., eds. *Southwestern Agriculture: Pre-Columbian to Modern*. College Station: Texas A & M University Press, 1982.

DeVorkin, David H. *Science with a Vengeance: How the Military Created the U.S. Space Sciences After World War II*. New York: Springer-Verlag, 1992.

Doel, Ronald E. *Solar System Astronomy in America: Communities, Patronage, and Interdisciplinary Research, 1920–1960*. New York: Cambridge University Press, 1996.

Douglass, A. E. *Climatic Cycles and Tree-Growth: A Study of the Annual Rings of Trees in Relation to Climate and Solar Activity*. Washington, D.C.: Carnegie Institution of Washington, 1919.

———. *Climatic Cycles and Tree-Growth*. Vol. 2: *A Study of the Annual Rings of Trees in Relation to Climate and Solar Activity*. Washington, D.C.: Carnegie Institution of Washington, 1928.

———. *Climatic Cycles and Tree-Growth*. Vol. 3: *A Study of Cycles*. Washington, D.C.: Carnegie Institution of Washington, 1936.

Dunning, Charles H., with Edward H. Peplow Jr. *Rock to Riches*. Phoenix: Southwest Publishing Company, Inc., 1959.

Edmondson, Frank K. *AURA and Its US National Observatories*. New York: Cambridge University Press, 1997.

Elliott, Melinda. *Great Excavations: Tales of Early Southwestern Archaeology, 1888–1939*. Santa Fe: School of American Research Press, 1995.

Engstrand, Iris H. W. *Spanish Scientists in the New World: The Eighteenth-Century Expeditions*. Seattle: University of Washington Press, 1981.

Etulain, Richard, ed. *Contemporary New Mexico, 1940–1990*. Albuquerque: University of New Mexico Press, 1994.

Evans, David S. and J. Derral Mulholland. *Big and Bright: A History of the McDonald Observatory*. Austin: University of Texas Press, 1986.

Fagette, Paul. *Digging for Dollars: American Archaeology and the New Deal*. Albuquerque: University of New Mexico Press, 1996.

Fernlund, Kevin J., ed. *The Cold War American West, 1945–1989*. Albuquerque: University of New Mexico Press, 1998.

Fewkes, Jesse Walter. *Hopi Katcinas*. New York: Dover Publications, 1985 (orig. pub. 1903).

Fireman, Janet R. *The Spanish Royal Corps of Engineers in the Western Borderlands: Instrument of Bourbon Reform, 1764 to 1815*. Glendale, California: Arthur H. Clark Company, 1977.

Fite, A. B., and Fabián García. "Pear variety test." New Mexico Agricultural Experiment Station, *Bulletin No. 165* (1928).

Fowler, Don D. *A Laboratory for Anthropology: Science and Romanticism in the American Southwest, 1846–1930*. Albuquerque: University of New Mexico Press, 2000.

Freeman, G. F. "Ripening Dates by Incubation." Arizona Agricultural Experiment Station, *Bulletin No. 66* (1911).

Fritts, Harold C. *Reconstructing Large-scale Climatic Patterns from Tree-Ring Data*. Tucson: University of Arizona Press, 1991.

Furman, Necah Stewart. *Sandia National Laboratories: The Postwar Decade*. Albuquerque: University of New Mexico Press, 1990.

García, Fabián. "Chile Culture." New Mexico Agricultural Experiment Station, *Bulletin No. 67* (1908).

——. "The effect of spring frosts on the peach crop; with cultural notes on the peach in New Mexico." New Mexico Agricultural Experiment Station, *Bulletin No. 30* (1899).

——. "Improved variety no. 9 of native chile." New Mexico Agricultural Experiment Station, *Bulletin No. 124* (1921).

——. "New Mexico Beans." New Mexico Agricultural Experiment Station, *Bulletin No. 105* (1917).

——. "Orchard notes: apricot, cherries, plums, quinces, figs. Pruning back the peach." New Mexico Agricultural Experiment Station, *Bulletin No. 39* (1901).

——. "Spraying orchards for the codling moth." New Mexico Agricultural Experiment Station, *Bulletin No. 41* (1902).

García, Fabián, and A. B. Fite. "Preliminary pecan experiments." New Mexico Agricultural Experiment Station, *Bulletin No. 145* (1925).

——. "Preliminary smudging experiments." New Mexico Agricultural Experiment Station, *Bulletin No. 134* (1922).

García, Fabián, and J. W. Rigney. "Winter protection of the vinifera grape." New Mexico Agricultural Experiment Station, *Bulletin No. 100* (1916).

Gerber, Stenehjem. *On the Home Front: The Cold War Legacy of the Hanford Nuclear Site.* Lincoln: University of Nebraska Press, 1992.

Goetzmann, William H. *Army Exploration in the American West, 1803–1863.* Lincoln: University of Nebraska Press, 1979 (orig. pub. 1959).

——. *Exploration and Empire: The Explorer and the Scientist in the Winning of the American West.* New York: Vintage Books, 1972 (orig. pub. 1966).

Goldsmith, Donald. *The Hunt for Life on Mars.* New York: Dutton, 1997.

Good, Gregory A., ed. *Sciences of the Earth: An Encyclopedia of Events, People, and Phenomena.* New York: Garland Publishing, Inc., 1998.

Gordon, Mary McDougall, ed. *Through Indian Country to California: John P. Sherburne's Diary of the Whipple Expedition, 1853–1854.* Stanford: Stanford University Press, 1988.

The Grand Central Tunnel Mining Company of New Mexico. Chicago: Culver, Page, Hoyne & Co., Printers, 1881.

Grobman, Arnold B. *The Changing Classroom: The Role of the Biological Sciences Curriculum Study.* Garden City, New York: Doubleday & Company, Inc., 1969.

Hall, R. Cargill. *Lunar Impact: A History of Project Ranger* [NASA SP-4210]. Washington, D.C.: National Aeronautics and Space Administration, 1977.

——. *Project Ranger: A Chronology* [JPL/HR-2]. Washington, D.C.: National Aeronautics and Space Administration, 1971.

Hansen, Harry, ed. *The World Almanac and Book of Facts.* New York: New York World-Telegram, 1950.

Harding, T. Swann. *Two Blades of Grass: A History of Scientific Development in the U.S. Department of Agriculture.* Norman: University of Oklahoma Press, 1947.

Hartmann, Gayle H., ed. *The Hodges Ruin: A Hohokam Community in the Tucson Basin.* Tucson: University of Arizona Press, 1978.

Haury, Emil W. *The Hohokam: Desert Farmers & Craftsmen. Excavations at Snaketown, 1964–1965*. Tucson: University of Arizona Press, 1976.

——. *Point of Pines, Arizona: A History of the University of Arizona Archaeological Field School*. Tucson: University of Arizona Press, 1989.

Hevley, Bruce, and John M. Findlay, eds. *The Atomic West*. Seattle: University of Washington Press, 1998.

Hinsley, Curtis M. *Savages and Scientists: The Smithsonian Institution and the Development of American Anthropology, 1846–1910*. Washington, D.C.: Smithsonian Institution Press, 1981.

Hinsley, Curtis M., and David R. Wilcox, eds. *The Southwest in the American Imagination: The Writings of Sylvester Baxter, 1881–1889*. Tucson: University of Arizona Press, 1996.

Hoddeson, Lillian, et al. *Critical Assembly: A Technical History of Los Alamos during the Oppenheimer Years, 1943–1945*. New York: Cambridge University Press, 1993.

Hopkins, Ernest J., and Alfred Thomas Jr. *The Arizona State University Story*. Phoenix: Southwest Publishing Co., Inc., 1960.

Hoyt, William Graves. *Coon Mountain Controversies: Meteor Crater and the Development of Impact Theory*. Tucson: University of Arizona Press, 1987.

——. *Lowell and Mars*. Tucson: University of Arizona Press, 1976.

——. *Planets X and Pluto*. Tucson: University of Arizona Press, 1980.

Hufbauer, Karl. *Exploring the Sun: Solar Science since Galileo*. Baltimore: Johns Hopkins University Press, 1991.

Hughes, Dorothy. *Pueblo on the Mesa: The First Fifty Years at the University of New Mexico*. Albuquerque: University of New Mexico Press, 1939.

Hurt, R. Douglas. *Indian Agriculture in America: Prehistory to the Present*. Lawrence: University Press of Kansas, 1987.

Istock, Conrad A., and Robert S. Hoffman, eds. *Storm over a Mountain Island: Conservation Biology and the Mt. Graham Affair*. Tucson: University of Arizona Press, 1995.

Jensen, Joan, and Darliss A. Miller, eds. *New Mexico Women: Intercultural Perspectives*. Albuquerque: University of New Mexico Press, 1986.

Jones, Billy M. *Health-Seekers in the Southwest: 1817–1900*. Norman: University of Oklahoma Press, 1967.

Jones, Fayette A. *Old Mining Camps of New Mexico, 1854–1904*. Santa Fe: Stagecoach Press, 1964.

King, C. J., et al. "Agricultural Investigations at the United States Field Station, Sacaton, Ariz., 1931–35." United States Department of Agriculture, *Circular No. 479* (July 1938).

Kofalk, Harriet. *No Woman Tenderfoot: Florence Merriam Bailey, Pioneer Naturalist*. College Station: Texas A & M University Press, 1989.

Koppes, Clayton R. *JPL and the American Space Program: A History of the Jet Propulsion Laboratory*. New Haven: Yale University Press, 1982.

Kropp, Simon F. *That All May Learn: New Mexico State University, 1886–1964*. Las Cruces: New Mexico State University, 1972.

Kunetka, James W. *City of Fire: Los Alamos and the Atomic Age, 1943–1945*. Rev. ed. Albuquerque: University of New Mexico Press, 1979.

Kunitz, Stephen J. *Disease Change and the Role of Medicine: The Navajo Experience*. Berkeley: University of California Press, 1983.

Lanham, Url. *The Bone Hunters*. New York: Columbia University Press, 1973.

Lankford, John. *American Astronomy: Community, Careers, and Power, 1859–1940*. Chicago: University of Chicago Press, 1997.

———, ed. *History of Astronomy: An Encyclopedia*. New York: Garland Publishing, 1997.

Lantz, Edith M. "Carotene and Ascorbic Acid Contents of Peppers." New Mexico Agricultural Experiment Station, *Bulletin 306* (1943).

———. "Effects of Canning and Drying on the Carotene and Ascorbic Acid Contents of Chile." New Mexico Agricultural Experiment Station, *Bulletin 327* (1946).

———. "Some Factors Affecting the Ascorbic Acid Content of Chile." New Mexico Agricultural Experiment Station, *Bulletin 324* (1945).

Larson, Edward J. *Summer for the Gods: The Scopes Trial and America's Continuing Debate Over Science and Religion*. New York: Basic Books, 1997.

———. *Trial and Error: The American Controversy Over Creation and Evolution*. New York: Oxford University Press, 1985.

Larson, Robert W. *New Mexico's Quest for Statehood, 1846–1912*. Albuquerque: University of New Mexico Press, 1968.

Lewis, Richard S. *Appointment on the Moon*. New York: Viking Press, 1968.

Limerick, Patricia Nelson. *The Legacy of Conquest: The Unbroken Past of the American West*. New York: W. W. Norton & Company, 1987.

Limerick, Patricia Nelson, et al., eds. *Trails: Toward a New Western History*. Lawrence: University Press of Kansas, 1991.

Lingenfelter, Richard E. *Steamboats on the Colorado River, 1852–1916*. Tucson: University of Arizona Press, 1978.

Linney, Charles E., and Fabián García. "Climate in relation to crop adaptation in New Mexico." New Mexico Agricultural Experiment Station, *Bulletin No. 113* (1918).

Loftin, John D. *Religion and Hopi Life in the Twentieth Century*. Bloomington: Indiana University Press, 1991.

Love, Frank. *Mining Camps and Ghost Towns: A History of Mining in Arizona and California Along the Lower Colorado*. Los Angeles: Westernlore Press, 1974.

Lowitt, Richard. *The New Deal and the West*. Bloomington: Indiana University Press, 1984.

Lucier, Paul. "Scientists and Swindlers: Coal, Oil, and Scientific Consulting in the American Industrial Revolution, 1830–1870." Ph.D. diss., Princeton University, 1994.

Luckingham, Bradford. *The Urban Southwest: A Profile History of Albuquerque—El Paso—Phoenix—Tucson*. El Paso: Texas Western Press, 1982.

McDougall, Walter A. . . . *the Heavens and the Earth: A Political History of the Space Age*. New York: Basic Books, 1985.

McGinnies, William G. *Discovering the Desert: Legacy of the Carnegie Desert Botanical Laboratory*. Tucson: University of Arizona Press, 1981.

Malville, J. McKim, and Claudia Putnam. *Prehistoric Astronomy in the Southwest*. Rev. ed. Boulder, Colorado: Johnson Books, 1993.

Martin, Douglas D. *The Lamp in the Desert: The Story of the University of Arizona*. Tucson: University of Arizona Press, 1960.

Meinig, D. W. *Southwest: Three Peoples in Geographical Change, 1600–1970*. New York: Oxford University Press, 1971.

Milner, Clyde A., II, et al. *The Oxford History of the American West*. New York: Oxford University Press, 1994.

Morrison, David. *Voyages to Saturn* [NASA SP-451]. Washington, D.C.: National Aeronautics and Space Administration, 1982.

Morrison, David, and Jane Samz. *Voyage to Jupiter* [NASA SP-439]. Washington, D.C.: National Aeronautics and Space Administration, 1980.

Nabhan, Gary Paul. *Gathering the Desert*. Tucson: University of Arizona Press, 1985.

Nash, Gerald D. *The American West in the Twentieth Century: A Short History of an Urban Oasis*. Albuquerque: University of New Mexico Press, 1977 (orig. pub. 1973).

———. *The American West Transformed: The Impact of the Second World War*. Bloomington: Indiana University Press, 1985.

———. *World War II and the West: Reshaping the Economy*. Lincoln: University of Nebraska Press, 1990.

Nash, Stephen Edward. *Time, Trees, and Prehistory: Tree-Ring Dating and the Development of North American Archaeology, 1914 to 1950*. Salt Lake City: University of Utah Press, 1999.

Norris, L. David, et al. *William H. Emory: Soldier-Scientist*. Tucson: University of Arizona Press, 1998.

Numbers, Ronald L. *The Creationists: The Evolution of Scientific Creationism*. New York: Alfred A. Knopf, 1992.

Ordway, Frederick I., III, and Mitchell R. Sharpe. *The Rocket Team*. New York: Thomas Y. Crowell, Publishers, 1979.

Osterbrock, Donald E. *Yerkes Observatory, 1892–1950: The Birth, Near Death, and Resurrection of a Scientific Research Institution*. Chicago: University of Chicago Press, 1997.

Parezo, Nancy J., ed. *Hidden Scholars: Women Anthropologists and the Native American Southwest*. Albuquerque: University of New Mexico Press, 1993.

Paterson, Kent Ian. *The Hot Empire of Chile*. Tempe, Arizona: Bilingual Press, 2000.

Peplow, Edward H. *History of Arizona*. New York: Lewis Historical Publishing Company, Inc., 1959.

Powell, John Wesley. *The Exploration of the Colorado River and its Canyons*. New York: Dover Publications, Inc., 1961 (orig. pub. 1895 as *Canyons of the Colorado*).

Project Y: The Los Alamos Story. Los Angeles: Tomash Publishers, 1983.

Putnam, William Lowell, et al. *The Explorers of Mars Hill: A Centennial History of Lowell Observatory, 1894–1994*. West Kennebunk, Maine: Phoenix Publishing, 1994.

Reid, Jefferson, and Stephanie Whittlesey. *The Archaeology of Ancient Arizona*. Tucson: University of Arizona Press, 1997.

Rhodes, Richard. *The Making of the Atomic Bomb*. New York: Simon and Schuster, 1986.

Rosenberg, Charles E. *No Other Gods: On Science and American Social Thought*. Rev. ed. Baltimore: Johns Hopkins University Press, 1997.

Rossiter, Margaret W. *Women Scientists in America: Struggles and Strategies to 1940*. Baltimore: Johns Hopkins University Press, 1982.

Rothman, Hal K. *On Rims & Ridges: The Los Alamos Area since 1880*. Lincoln: University of Nebraska Press, 1992.

——. *Preserving Different Pasts: The American National Monuments*. Urbana: University of Illinois Press, 1989.

Schackel, Sandra. *Social Housekeepers: Women Shaping Public Policy in New Mexico, 1920–1940*. Albuquerque: University of New Mexico Press, 1992.

Schwantes, Carlos A. *Vision and Enterprise: Exploring the History of Phelps Dodge Corporation*. Tucson: University of Arizona Press, 2000.

Sheridan, Thomas E. *Arizona: A History*. Tucson: University of Arizona Press, 1995.

Sherman, James E. and Barbara H. *Ghost Towns and Mining Camps of New Mexico*. Norman: University of Oklahoma Press, 1975.

Shirley, Donna. *Managing Martians*. New York: Broadway Books, 1998.

Shreve, Edith. *The Daily March of Transpiration in a Desert Perennial*. Washington, D.C.: Carnegie Institution, 1914.

Silliman, Benjamin. *The Rio Grande Gold Gravels*. Omaha: Henry Gibson, Herald Publishing House, 1880.

Simmons, Marc. *Albuquerque: A Narrative History*. Albuquerque: University of New Mexico Press, 1982.

Smith, Margaret C., et al. "The Comparative Vitamin C Values of Arizona Citrus Fruits of Different Varieties and Sizes When Prepared for Consumption in Several Different Ways." Arizona Agricultural Experiment Station, *Mimeographed Report 62* (1944).

Snead, James E. *Ruins and Rivals: The Making of Southwestern Archaeology*. Tucson: University of Arizona Press, 2001.

Sonnichsen, C. L. *Pass of the North: Four Centuries on the Rio Grande*. El Paso: Texas Western Press, 1980.

Spence, Clark C. *British Investments and the American Mining Frontier, 1860–1901*. Ithaca: Cornell University Press, 1958.

——. *Mining Engineers in the American West: The Lace-Boot Brigade, 1849–1933*. Moscow: University of Idaho Press, 1993 (orig. pub. 1970).

Spidle, Jake W., Jr. *Doctors of Medicine in New Mexico: A History of Health and Medical Practice, 1886–1986*. Albuquerque: University of New Mexico Press, 1986.

Stegner, Wallace. *Beyond the Hundredth Meridian: John Wesley Powell and the Second Opening of the West*. Houghton Mifflin Co., 1954.

Strauss, David. *Percival Lowell: The Culture and Science of a Boston Brahmin*. Cambridge: Harvard University Press, 2001.

Swick, Mike E. *Cultural Resource Survey of Turquoise Hill, Santa Fe County, New Mexico*. Santa Fe: New Mexico Abandoned Mine Land Bureau, 1995.

Swingle, W. T. "The Date Palm and Its Utilization in the Southwestern States." U.S. Department of Agriculture, Bureau of Plant Industry, *Bulletin No. 53* (1904).

Szasz, Ferenc Morton. *The Day the Sun Rose Twice: The Story of the Trinity Site Nuclear Explosion, July 16, 1945*. Albuquerque: University of New Mexico Press, 1984.

Tatarewicz, Joseph N. *Space Technology and Planetary Astronomy*. Bloomington: Indiana University Press, 1990.

Toumey, Christopher P. *God's Own Scientists: Creationists in a Secular World*. New Brunswick, New Jersey: Rutgers University Press, 1994.

Toumey, J. W. "The Date Palm." Arizona Agricultural Experiment Station, *Bulletin No. 29* (1898).

Trennert, Robert A. *White Man's Medicine: Government Doctors and the Navajo, 1863–1955*. Albuquerque: University of New Mexico Press, 1998.

Twitchell, Emerson. *The Leading Facts of New Mexican History*. Cedar Rapids, Iowa: The Torch Press, 1911.

Vinson, A. E. "Chemistry and Ripening of the Date." Arizona Agricultural Experiment Station, *Bulletin No. 66* (1911).

Visher, Stephen S. *Scientists Starred 1903–1943 in "American Men of Science"*. New York: Arno Press, 1975 (orig. pub. 1947).

Wagoner, Jay J. *Arizona Territory, 1863–1912: A Political History*. Tucson: University of Arizona Press, 1970.

Waters, Frank. *Book of the Hopi*. New York: Ballantine Books, 1977 (orig. pub. 1963).

Webb, George E. *The Evolution Controversy in America*. Lexington: University Press of Kentucky, 1994.

——. *Tree Rings and Telescopes: The Scientific Career of A. E. Douglass*. Tucson: University of Arizona Press, 1983.

Weber, David J. *The Spanish Frontier in North America*. New Haven: Yale University Press, 1992.

Welsh, Michael E. *U.S. Army Corps of Engineers, Albuquerque District, 1935–1985*. Rev. ed. Albuquerque: University of New Mexico Press, 1987.

Whitaker, Ewen A. *The University of Arizona's Lunar and Planetary Laboratory: Its Founding and Early Years*. Tucson: University of Arizona, 1985.

White, Gerald T. *Scientists in Conflict: The Beginnings of the Oil Industry in California*. San Marino, California: The Huntington Library, 1968.

White, Richard. *"It's Your Misfortune and None of My Own": A History of the American West*. Norman: University of Oklahoma Press, 1991.

Wilhelm, Richard David. "A Chronology and Analysis of Regulatory Actions Relating to the Teaching of Evolution in Public Schools." Ph.D. diss., University of Texas at Austin, 1978.

Wilhelms, Don E. *To a Rocky Moon: A Geologist's History of Lunar Exploration*. Tucson: University of Arizona Press, 1993.

Wilson, Leonard G., ed. *Benjamin Silliman and His Circle: Studies on the Influence of*

Benjamin Silliman on Science in America. New York: Science History Publications, 1979.

Woodbury, Richard B. *Sixty Years of Southwestern Archaeology: A History of the Pecos Conference*. Albuquerque: University of New Mexico Press, 1993.

Worster, Donald. *A River Running West: The Life of John Wesley Powell*. New York: Oxford University Press, 2001.

The WPA Guide to 1930s Arizona. Tucson: University of Arizona Press, 1989 (orig. pub. 1940).

Articles

Abbot, C. G. "The Relation of the Sun-Spot Cycle to Meteorology." *Monthly Weather Review* 30 (April 1902): 178–181.

Abt, Helmut A. "Award of the Bruce Gold Medal to Professor Willem J. Luyten." Astronomical Society of the Pacific, *Publications* 80 (June 1968): 247–251.

Albee, A. L., et al. "Mars Global Surveyor Mission: Overview and Status." *Science* 279 (13 March 1998): 1671–1672.

Alexander, Thomas G. "From Rule of Thumb to Scientific Range Management: The Case of the Intermountain Region of the Forest Service." *Western Historical Quarterly* 18 (October 1987): 409–428.

Allen, J. A. "The Masked bob-white (Colinus Ridgwayi) of Arizona, and its Allies." American Museum of Natural History, *Bulletin* 1 (July 1886): 273–290.

———. "On a Collection of Mammals from Arizona and Mexico, Made by Mr. W. W. Price, with Field Notes by the Collector." American Museum of Natural History, *Bulletin* 7 (June 1895): 193–258.

Anderson, Christopher. "Weapons Labs in a New World." *Science* 262 (8 October 1993): 168–171.

Anderson, Kurt S. J. "To the Edge of the Universe: The Apache Point Observatory." *New Mexico Journal of Science* 35 (November 1995): 6–20.

Atherton, Lewis. "The Mining Promoter in the Trans-Mississippi West." *Western Historical Quarterly* 1 (January 1970): 35–50.

Baatz, Simon. "'Squinting at Silliman': Scientific Periodicals in the Early American Republic." *Isis* 82 (June 1991): 223–244.

Bailey, Florence Merriam. "Additional Notes on the Birds of the Upper Pecos." *Auk* 21 (July 1904): 349–363.

———. "Additions to Mitchell's List of the Summer Birds of San Miguel County, New Mexico." *Auk* 21 (October 1904): 443–449.

———. "A Drop of Four Thousand Feet." *Auk* 28 (April 1911): 219–225.

———. "Notable Migrants Not Seen at Our Arizona Bird Table." *Auk* 40 (July 1923): 393–409.

Bartlett, Richard A. "Scientific Exploration of the American West, 1865–1900." In *North American Exploration*, vol. 3, edited by John Logan Allen. Lincoln: University of Nebraska Press, 1997.

Beebe, Reta, and Nancy Chanover. "Atmospheres of the Giant Planets." *New Mexico Journal of Science* 35 (November 1995): 86–101.

Bigelow, Frank H. "Studies on the Diurnal Periods in the Lower Strata of the Atmosphere." *Monthly Weather Review* 33 (July 1905): 292–295.

Binder, Alan B. "Lunar Prospector: An Overview." *Science* 281 (4 September 1998): 1475–1476.

Brandt, John C., and Ray A. Williamson. "Rock Art Representations of the A.D. 1054 Supernova: A Progress Report." In *Native American Astronomy*, edited by Anthony F. Aveni. Austin: University of Texas Press, 1977.

Brandt, John C., et al. "Possible Rock Art Records of the Crab Nebula Supernova in the Western United States." In *Archaeoastronomy in Pre-Columbian America*, edited by Anthony F. Aveni. Austin: University of Texas Press, 1975.

Broadfoot, A. L., et al. "Ultraviolet Spectrometer Observations of Neptune and Triton." *Science* 246 (15 December 1989): 1459–1466.

Brown, Barnum. "The Cretaceous Ojo Alamo Beds of New Mexico with Description of the New Dinosaur Genus *Kritosaurus*." American Museum of Natural History, *Bulletin* 28 (July 1910): 267–274.

Brown, Herbert. "Arizona Bird Notes." *Auk* 20 (January 1903): 43–50.

———. "A Band-tailed Hawk's Nest." *Auk* 18 (October 1901): 392–393.

———. "Bendire's Thrasher." *Auk* 18 (July 1901): 225–231.

———. "Masked Bob-white (*Colinus Ridgwayi*)." *Auk* 21 (April 1904): 209–213.

Carpenter, Edwin F. "A Cluster of Extra-Galactic Nebulae in Cancer." Astronomical Society of the Pacific, *Publications* 43 (August 1931): 247–254.

———. "The Distribution of Color in Two Extra-Galactic Nebulae." Astronomical Society of the Pacific, *Publications* 43 (August 1931): 294–295.

———. "Note on the Absorption of Radiation in Space." American Astronomical Society, *Publications* 7 (1933): 25.

———. "Spectra of Binary Galaxies Photographed with an Echelle-Type Nebular Spectrograph." Astronomical Society of the Pacific, *Publications* 69 (October 1957): 386–388.

———. "U Cephei: An Anomalous Spectrographic Result." *Astrophysical Journal* 72 (November 1930): 205–220.

Carpenter, Edwin F., et al. "Determination of Color Classes for 204 Stars of Large Proper Motion." *Astrophysical Journal* 116 (November 1952): 587–591.

Cattell, J. McKeen. "The Origin and Distribution of Scientific Men." *Science* 66 (25 November 1927): 513–516.

Chaikin, Andrew. "Hard Landings." *Air & Space Smithsonian* 12 (June/July 1997): 48–55.

Chapin, Seymour. "Garrett and Pressurized Flight: A Business Built on Thin Air." *Pacific Historical Review* 35 (August 1966): 329–343.

Christensen, P. R. "Results from the Mars Global Surveyor Thermal Emission Spectrometer." *Science* 279 (13 March 1998): 1692–1698.

Church, Samuel Harden. "Scott, Thomas Alexander." In *Dictionary of American Biogra-*

phy, edited by Dumas Malone. New York: Charles Scribner's Sons, 1963 (orig. pub. 1935).

Clough, H. W. "Synchronous Variations in Solar and Terrestrial Phenomena." *Astrophysical Journal* 22 (July 1905): 42–75.

Cockerell, T.D.A. "Notes on New Mexican Flowers and Their Insect Visitors." *Botanical Gazette* 24 (August 1897): 104–197.

Colley, Charles C. "Arizona, Cradle of the American Date Growing Industry, 1890–1916." *Southern California Quarterly* 53 (March 1971): 55–66.

Conrad, David E. "The Whipple Expedition in Arizona, 1853–1854." *Arizona and the West* 11 (Summer 1969): 147–178.

Cope, Edward D. "First Addition to the Fauna of the Puerco Eocene." American Philosophical Society, *Proceedings* 20 (1883): 545–563.

——. "On some Mammalia of the Lowest Eocene beds of New Mexico." American Philosophical Society, *Proceedings* 19 (1880–1881): 484–495.

——. "On the REPTILIA and BATRACHIA of the Sonoran Province of the Nearctic Region." Academy of Natural Sciences of Philadelphia, *Proceedings* 18 (October 1866): 300–314.

——. "The Permian Formation of New Mexico." *American Naturalist* 15 (December 1881): 1020–1021.

——. "Synopsis of the Vertebrata Fauna of the Puerco Series." American Philosophical Society, *Transactions* 16 (1890): 298–361.

——. "Synopsis of the Vertebrata of the Puerco Eocene epoch." American Philosophical Society, *Proceedings* 20 (1883): 461–471.

Cotter, Donald J. "The Scientific Contribution of New Mexico to the Chile Pepper." In *Southwestern Agriculture: Pre-Columbian to Modern*, edited by Henry C. Dethloff and Irvin M. May Jr. College Station: Texas A & M University Press, 1982.

Coues, Elliott. "List of the Birds of Fort Whipple, Arizona." Academy of Natural Sciences of Philadelphia, *Proceedings* 18 (March 1866): 39–100.

——. "Notes on a Collection of Mammals from Arizona." Academy of Natural Sciences of Philadelphia, *Proceedings* 19 (November 1867): 133–136.

——. "Notes on various birds observed at Fort Whipple, Ariz." *Ibis* 2d ser., 1 (October 1865): 535–538.

——. "Ornithology of a Prairie-Journey, and Notes on the Birds of Arizona." *Ibis* 2d ser., 1 (April 1865): 157–165.

——. "The Quadrupeds of Arizona." *American Naturalist* 1 (August 1867): 281–292; (September 1867): 351–363; (October 1867): 393–400; (December 1867): 531–541.

Crosswhite, Frank S. " 'J. G. Lemmon & Wife,' Plant Explorers in Arizona, California, and Nevada." *Desert Plants* 1 (August 1979): 12–21.

Cruikshank, Dale P. "Gerard Peter Kuiper." In National Academy of Sciences, *Biographical Memoirs* 62 (1993): 259–295.

"Darwin's Brush With Racism." *Science* 292 (18 May 2001): 1295.

"DOE won't run tests at WIPP." *ENR* 231 (1 November 1993): 7.

Douglass, A. E. "Drawings of Comet *a*, 1910." *Popular Astronomy* 18 (March 1910): 162–163.

———. "Evidences of Cycles in Tree Ring Records." National Academy of Sciences, *Proceedings* 19 (1933): 350–360.

———. "Historical Address." *The Inaugural Bulletin* (Tucson: University of Arizona, 1923): 54–59.

———. "An Optical Periodograph." *Astrophysical Journal* 41 (April 1915): 173–186.

———. "A Photographic Periodogram of the Sun-Spot Numbers." *Astrophysical Journal* 40 (October 1914): 326–331.

———. "The University of Arizona Eclipse Expedition. Port Libertad, Sonora, Mexico, September 10, 1923." Astronomical Society of the Pacific, *Publications* 36 (August 1924): 170–184.

———. "Weather Cycles in the Growth of Big Trees." *Monthly Weather Review* 37 (June 1909): 225–237.

Edwords, Frederick. "Creation-Evolution Debates: Who's Winning Them Now?" *Creation/Evolution* 3 (Spring 1982): 30–42.

Ellis, Florence Hawley. "A Thousand Years of the Pueblo Sun-Moon-Star Calendar." In *Archaeoastronomy in Pre-Columbian America*, edited by Anthony F. Aveni. Austin: University of Texas Press, 1975.

Emory, Deborah Carley. "Running the Line: Men, Maps, Science, and Art of the United States and Mexico Boundary Survey, 1849–1856." *New Mexico Historical Review* 75 (April 2000): 221–265.

Endlich, F. M. "The Mining Regions of Southern New Mexico." *American Naturalist* 17 (February 1883): 149–157.

Feldman, W. C., et al. "Fluxes of Fast and Epithermal Neutrons from Lunar Prospector: Evidence for Water Ice at the Lunar Poles." *Science* 281 (4 September 1998): 1496–1500.

Ferber, Dan. "GM Crops in the Cross Hairs." *Science* 286 (26 November 1999): 1662–1666.

Finch, A. H., and W. T. McGeorge. "Studies of Grapefruit Fertilization in Arizona." American Society for Horticultural Science, *Proceedings* 37 (May 1940): 62–67.

Finley, David G. "Radio Astronomy in New Mexico: The VLA and VLBA." *New Mexico Journal of Science* 35 (November 1995): 21–26.

Fisher, Allan C., Jr. "Aviation Medicine on the Threshold of Space." *National Geographic Magazine* 108 (August 1955): 241–278.

Fitch, Walter S. "The Cluster-Type Variable VZ Cancri." *Astrophysical Journal* 121 (May 1955): 690–710.

———. "The Light-Variation of CC Andromedae." *Astrophysical Journal* 132 (November 1960): 701–715.

———. "Photoelectric Observations of the Cluster-Type Variables EH Librae and DY Herculis." *Astronomical Journal* 62 (May 1957): 108–111.

Flam, Faye. "Is There Life after the SSC?" *Science* 262 (29 October 1993): 644–647.

Gilruth, Robert R. "The Making of an Astronaut." *National Geographic Magazine* 127 (January 1965): 122–144.

Golombek, Matthew P. "A Message from Warmer Times." *Science* 283 (5 March 1999): 1470–1471.

Goodwin, Irwin. "After Agonizing Death in the Family, Particle Physics Faces Grim Future." *Physics Today* 47 (February 1994): 87–91.

———. "Amazing Race: The SSC Contest Generated Disorder and Discord." *Physics Today* 41 (May 1988): 69–74.

———. "To Replace AT&T at Sandia, DOE Picks Martin Marietta." *Physics Today* 46 (September 1993): 53–54.

Gore, Rick. "Neptune: Voyager's Last Picture Show." *National Geographic Magazine* 178 (August 1990): 35–47.

———. "Uranus: Voyager Visits a Dark Planet." *National Geographic Magazine* 170 (August 1986): 179–194.

———. "Voyager 1 at Saturn: Riddles of the Rings." *National Geographic Magazine* 160 (July 1981): 3–31.

"Green Light for Mount Graham Telescope?" *Science* 272 (3 May 1996): 637.

Handschin, C. H. "The Percentage of Women Teachers in State Colleges and Universities." *Science* 35 (12 January 1912): 55–57.

Hewitt, Harry P. "The Mexican Boundary Survey Team: Pedro García Conde in California." *Western Historical Quarterly* 21 (May 1990): 171–196.

Hilgeman, R. H., and J. G. Smith. "Changes in Invert Sugar and Sucrose During Ripening of Arizona Grapefruit." American Society for Horticultural Science, *Proceedings* 37 (May 1940): 535–538.

Hinton, Harwood P. "Frontier Speculation: A Study of the Walker Mining Districts." *Pacific Historical Review* 29 (August 1960): 245–255.

Holliday, Clyde T. "Seeing the Earth from 80 Miles Up." *National Geographic Magazine* 98 (October 1950): 511–528.

Hoyt, William Graves. "Vesto Melvin Slipher." In National Academy of Sciences, *Biographical Memoirs* 52 (1980): 411–449.

Hubble, Edwin, and Milton L. Humason. "The Velocity-Distance Relation Among Extra-Galactic Nebulae." *Astrophysical Journal* 74 (July 1931): 43–80.

Hviid, S. F., et al. "Magnetic Properties Experiments on the Mars Pathfinder Lander: Preliminary Results." *Science* 278 (5 December 1997): 1768–1770.

Irion, Robert. "Lunar Prospector Probes Moon's Core Mysteries." *Science* 281 (4 September 1998): 1423–1424.

James, Joseph F. "Botanical Notes from Tucson." *American Naturalist* 15 (December 1881): 978–987.

Jeffrey, Duane E. "Seers, Savants and Evolution: The Uncomfortable Interface." *Dialogue: A Journal of Mormon Thought* 8 (Autumn/Winter 1974): 41–75.

Jensen, Joan. "Canning Comes to New Mexico: Women and the Agricultural Extension Service, 1914–1919." In *New Mexico Women: Intercultural Perspectives*, edited by

Joan M. Jensen and Darlis A. Miller. Albuquerque: University of New Mexico Press, 1986.

Jepson, Willis L. "Lemmon, John Gill." In *Dictionary of American Biography*, edited by Dumas Malone. New York: Charles Scribner's Sons, 1961 (orig. pub. 1933).

Johnson, Judith R. "John Weinzirl: A Personal Search for the Conquest of Tuberculosis." *New Mexico Historical Review* 63 (April 1988): 141–155.

Jones, Oakah L., Jr. "Spanish Penetrations to the North of New Spain." In *North American Exploration*, vol. 2, edited by John Logan Allen. Lincoln: University of Nebraska Press, 1997.

Jones, Winston W., et al. "A Note on Ascorbic Acid: Nitrogen Relationships in Grapefruit." *Science* 99 (4 February 1944): 103–104.

Jones, Winston W., et al. "The Relation of Nitrogen Absorption to Nitrogen Content of Fruit and Leaves in Citrus." American Society for Horticultural Science, *Proceedings* 45 (November 1944): 1–4.

Kaiser, Jocelyn. "Plan Would Shut Kitt Peak Facilities." *Science* 272 (3 May 1996): 641.

Kerr, Richard A. "Ancient Life on Mars?" *Science* 273 (16 August 1966): 864–866.

——. "Cheapest Mission Finds Moon's Frozen Water." *Science* 281 (4 September 1998): 1628–1629.

——. "Gambling On a Martian Landing Site." *Science* 272 (19 April 1995): 347–348.

——. "Pathfinder Tells a Geologic Tale With One Starring Role." *Science* 279 (9 January 1998): 175.

——. "Possible Glimpse of Earth like Geology In Mars Rock." *Science* 277 (1 August 1997): 638–639.

——. "Rocky Mix Suggests Wet Early Mars." *Science* 278 (17 October 1997): 380.

Kessell, John L. "'To See Such Marvels With My Own Eyes': Spanish Exploration in the Western Borderlands." *Montana: The Magazine of Western History* 41 (Autumn 1991): 68–75.

Koshland, Daniel E., Jr. "Dealing in Hot Property." *Science* 232 (27 June 1986): 1585.

Kuslan, Louis I. "Benjamin Silliman, Jr.: The Second Silliman." In *Benjamin Silliman and His Circle: Studies on the Influence of Benjamin Silliman on Science in America*, edited by Leonard G. Wilson. New York: Science History Publications, 1979.

Langley, S. P. "On a Possible Variation of the Solar Radiation and its Probable Effect on Terrestrial Temperatures." *Astrophysical Journal* 19 (June 1904): 305–321.

Lankford, John, and Rickey L. Slavings. "Gender and Science: Women in American Astronomy, 1859–1940." *Physics Today* 43 (March 1990): 58–65.

Lawler, Andrew. "Finding Puts Mars Exploration on Front Burner." *Science* 273 (16 August 1996): 865.

——. "Small Missions Lift Planetary Science." *Science* 277 (12 September 1997): 1596–1598.

Lee, Sally J., and Jeffrey P. Brown. "Women at New Mexico State University: The Early Years, 1888–1920." *New Mexico Historical Review* 64 (January 1989): 77–93.

Lemmon, J. G. "A Botanical Wedding Trip." *Californian* 4 (December 1881): 517–525.

Lisonbee, Lorenzo. "Thwarting the Anti-Evolution Movement in Arizona." *Science Teacher* 32 (February 1965): 35–37.

Lorbiecki, Marybeth. "The Land Makes the Man: New Mexico's Influence on the Conservationist Aldo Leopold." *New Mexico Historical Review* 73 (July 1998): 235–252.

Lucier, Paul. "Commercial Interests and Scientific Disinterestedness: Consulting Geologists in Antebellum America." *Isis* 86 (June 1995): 245–267.

Luyten, Willem J. "The Search for White Dwarfs." *Astronomical Journal* 55 (April 1950): 86–89.

Luyten, Willem J., and Edwin F. Carpenter. "First Report on a Systematic Survey for Faint Blue Stars." *Astronomical Journal* 60 (December 1955): 429–431.

McCluskey, Stephen C. "Historical Archaeoastronomy: The Hopi Example." In *Archaeoastronomy in the New World*, edited by Anthony F. Aveni. Cambridge: Cambridge University Press, 1982.

McDonald, Kim A. "Researchers Battle for Access to a 9,300-Year-Old Skeleton." *Chronicle of Higher Education* 44 (22 May 1998): A18–A19, A22.

Malin, M. C. "Early Views of the Martian Surface from the Mars Orbiter Camera of *Mars Global Surveyor*." *Science* 279 (13 March 1998): 1681–1685.

Marshall, Eliot. "The Supercollider Sweepstakes." *Science* 237 (11 September 1987): 1288.

Martin, Buddy, et al. "The New Ground-Based Optical Telescopes." *Physics Today* 44 (March 1991): 22–30.

Martin, W. E., et al. "Grapefruit Storage Studies in Arizona." American Society for Horticultural Science, *Proceedings* 37 (May 1940): 529–534.

Masse, W. Bruce. "Prehistoric Irrigation Systems in the Salt River Valley, Arizona." *Science* 214 (23 October 1981): 408–415.

Matijevic, J. "Autonomous Navigation and the *Sojourner* Microrover." *Science* 280 (17 April 1998): 454–455.

Matsumura, Molleen. "Arizona: Another Sunbelt State Axes Evolution." National Center for Science Education, *Reports* 17 (July/August 1997): 4.

———. "Does Arizona Charter School Teach Creationism?" National Center for Science Education, *Reports* 16 (Fall 1996): 6.

Matthew, W. D. "A Revision of the Puerco Fauna." American Museum of Natural History, *Bulletin* 9 (November 1897): 259–323.

Mearns, Edgar A. "Description of a New Species of Weasel, and a New Subspecies of the Gray Fox, from Arizona." American Museum of Natural History, *Bulletin* 3 (February 1891): 234–238.

———. "Description of a rare Squirrel, new to the Territory of Arizona." American Museum of Natural History, *Bulletin* 1 (July 1886): 197–207.

———. "Description of supposed New Species and Subspecies of Mammals, from Arizona." American Museum of Natural History, *Bulletin* 2 (February 1890): 277–307.

———. "Descriptions of a New Species and Three New Subspecies of Birds from Arizona." *Auk* 7 (July 1890): 243–251.

———. "Descriptions of Three New Forms of Pocket-Mice from the Mexican Border of the United States." American Museum of Natural History, *Bulletin* 10 (August 1898): 299–302.

——. "Notes on the Otter (Lutra canadensis) and Skunks (Genera Spilogale and Mephitis) of Arizona." American Museum of Natural History, *Bulletin* 3 (February 1891): 252–262.

——. "Observations on the Avifauna of Portions of Arizona." *Auk* 7 (January 1890): 45–55; (July 1890): 251–264.

"Merriam's 'Results of a Biological Survey of the San Francisco Mountain Region and desert of the Little Colorado, Arizona'." *Auk* 8 (January 1891): 95–98.

Mervis, Jeffrey. "Red Squirrels 2, Astronomers 0." *Science* 268 (5 May 1995): 630.

Minckley, W. L. "Frederic Morton Chamberlain's 1904 Survey of Arizona Fishes, with Annotations." *Journal of the Southwest* 41 (Summer 1999): 177–237.

Morell, Virginia. "Kennewick Man's Trials Continue." *Science* 280 (10 April 1998): 190–192.

"Mt. Graham Site Faces Opposition." American Astronomical Society, *Newsletter* 50 (June 1990): 1.

Neilson, Ronald P. "High Resolution Climatic Analysis and Southwest Biogeography." *Science* 232 (4 April 1986): 27–34.

"New Mexico Funds New NRAO Building." American Astronomical Society, *Newsletter* 29 (March 1986): 8–9.

Newcott, William R. "Return to Mars." *National Geographic Magazine* 194 (August 1998): 2–29.

"Oklahoma Lawmakers Take a Shot at Darwin." *Science* 288 (21 April 2000): 431.

"The Oldest Tertiary Mammalia." *American Naturalist* 19 (April 1885): 385–387.

Osborn, Henry Fairfield, and Charles Earle. "Fossil Mammals of the Puerco Beds. Collection of 1892." American Museum of Natural History, *Bulletin* 7 (February 1895): 1–70.

Palmer, Edward. "Plants Used by the Indians of the United States." *American Naturalist* 12 (September 1878): 593–606; (October 1878): 646–655.

Porco, Carolyn C. "An Explanation for Neptune's Ring Arcs." *Science* 253 (30 August 1991): 995–1001.

"The Progress of the Ungulates in Tertiary Time." *American Naturalist* 17 (October 1883): 1055–1057.

Proulx, E. A. "Some Like Them Hot." *Horticulture* 63 (January 1985): 46–54.

Reinhartz, Dennis, and Oakah L. Jones. *"Hacia el Norte!* The Spanish *Entrada* into North America, 1513–1549." In *North American Exploration*, vol. 1, edited by John Logan Allen. Lincoln: University of Nebraska Press, 1997.

Rice, Virginia E. "The Arizona Agricultural Experiment Station: A History to 1917." *Arizona and the West* 20 (Summer 1978): 123–140.

Richmond, Charles W. "In Memorium: Edgar Alexander Mearns." *Auk* 35 (January 1918): 1–18.

Roach, F. E. "A Composite Light-Curve of Eros." *Astrophysical Journal* 95 (March 1942): 310–313.

Roach, F. E., and Laurence G. Stoddard. "A Photoelectric Light-Curve of Eros." *Astrophysical Journal* 88 (October 1938): 305–312.

Roehm, Gladys Hartley. "The Vitamin B and G Content of Arizona-Grown Grapefruit and Broccoli." *Journal of Home Economics* 27 (December 1935): 663–666.

Rover Team. "Characterization of the Martian Surface Deposits by the *Mars Pathfinder* Rover, Sojourner." *Science* 278 (5 December 1997): 1765–1768.

Saltzman, Martin D. "The Art of Distillation and the Dawn of the Hydrocarbon Society." *Bulletin for the History of Chemistry* 24 (1999): 53–60.

Schlesinger, William H., et al. "Biological Feedbacks in Global Desertification." *Science* 247 (2 March 1990): 1043–1048.

Schofield, J. T. "The *Mars Pathfinder* Atmospheric Structure Investigation/Meteorology (ASI/MET) Experiment." *Science* 278 (5 December 1997): 1752–1758.

Schroeder, Albert H. "The Cerillos Mining Area." In *Archaeology and History of Santa Fe Country*. Santa Fe: New Mexico Geological Society, 1997.

Scott, Ed. "Was There Life in the Martian Meteorite?" *Mercury* 27 (September–October 1998): 8–9.

Scott, Eugenie C. "Not (Just) in Kansas Anymore." *Science* 288 (5 May 2000): 813, 815.

Shreve, Edith B. "Investigations on the Imbibition of Water by Gelatine." *Science* 48 (27 September 1918): 324–327.

———. "The Relation of Transpiration to Evaporation from Artificial Surfaces." *American Journal of Botany* 27 (October 1940): 707.

———. "Seasonal Changes in the Water Relations of Desert Plants." *Ecology* 4 (July 1923): 266–292.

Silliman, Benjamin. "The Mineral Regions of Southern New Mexico." American Institute of Mining Engineers, *Transactions* 10 (1881–1882): 424–444.

———. "Mineralogical Notes." *American Journal of Science* 3d. ser., 22 (September 1881): 198–205.

———. "On some of the Mining Districts of Arizona near the Rio Colorado, with remarks on the Climate, &c." *American Journal of Science* 2d ser., 41 (May 1866): 289–308.

———. "Turquois of New Mexico." *American Journal of Science* 3d ser., 22 (July 1881): 67–71.

Sinclair, W. J., and Walter Granger. "Paleocene Deposits of the San Juan Basin, New Mexico." American Museum of Natural History, *Bulletin* 33 (June 1914): 297–316.

Smith, Bradford A. "Voyage of the Century." *National Geographic Magazine* 178 (August 1990): 48–65.

Smith, Bradford A., et al. "The Jupiter System Through the Eyes of Voyager 1." *Science* 204 (1 June 1979): 956–969.

Smith, Bradford A., et al. "A New Look at the Saturn System: The Voyager 2 Images." *Science* 215 (29 January 1982): 504–537.

Smith, Bradford A., et al. "Voyager 2 at Neptune: Imaging Science Results." *Science* 246 (15 December 1989): 1422–1449.

Smith, Bradford A., et al. "Voyager 2 in the Uranian System: Imaging Science Results." *Science* 233 (4 July 1986): 43–64.

Smith, P. H., et al. "Results from the Mars Pathfinder Camera." *Science* 278 (5 December 1997): 1758–1765.

Sofaer, Anna, et al. "Lunar Markings on Fajada Butte, Chaco Canyon, New Mexico." In *Archaeoastronomy in the New World*, edited by Anthony F. Aveni. Cambridge: Cambridge University Press, 1982.

Sofaer, Anna, et al. "A Unique Solar Marking Construct." *Science* 206 (19 October 1979): 283–291.

Spence, Clark C. "The Janin Brothers: Mining Engineers." *Mining History Journal* 3 (1996): 76–82.

Stiles, Lori. "UA Scientists Show Mars to the World." *Arizona Alumnus* 75 (Fall 1997): 12–13.

Stocking, George W., Jr. "The Santa Fe Style in American Anthropology: Regional Interest, Academic Initiative, and Philanthropic Policy in the First Two Decades of the Laboratory of Anthropology, Inc." *Journal of the History of the Behavioral Sciences* 18 (1982): 3–19.

Stone, E. C., and E. D. Miner. "Voyager 2 Encounter with the Saturnian System." *Science* 215 (29 January 1982): 499–504.

———. "The Voyager 2 Encounter with the Uranian System." *Science* 233 (4 July 1986): 39–43.

Szasz, Ferenc M. "The Cultures of Modern New Mexico, 1940–1990." In *Contemporary New Mexico, 1940–1990*, edited by Richard W. Etulain. Albuquerque: University of New Mexico Press, 1994.

Terrile, Richard J., and Reta F. Beebe. "Summary of Historical Data: Interpretation of the Pioneer and Voyager Cloud Configurations in a Time-Dependent Framework." *Science* 204 (1 June 1979): 948–951.

Thompson, Gerald. "'Is there a Gold Field East of the Colorado?' The La Paz Rush of 1862." *Southern California Quarterly* 67 (Winter 1985): 345–363.

Toumey, J. W. "A bit of the flora of Central Arizona." *Botanical Gazette* 17 (May 1892): 162–164.

Travis, John. "Scopes and Squirrels Return to Court." *Science* 265 (2 September 1994): 1356.

"Tucson Gets Experiment Station." *American Forests and Forest Life* 36 (July 1930): 467.

Very, Frank W. "The Variation of Solar Radiation." *Astrophysical Journal* 7 (April 1898): 255–272.

Vinson, A. E. "The Stimulation of Premature Ripening by Chemical Means." American Chemical Society, *Journal* 32 (February 1910): 208–212.

Waldrop, M. Mitchell. "The Long, Sad Saga of Mount Graham." *Science* 248 (22 June 1990): 1479–1481.

———. "The New Art of Telescope Making." *Science* 234 (19 December 1986): 1495–1497.

———. "Radio Astronomy's Crumbling Showpiece." *Science* 253 (19 July 1991): 268–269.

———. "Troubled Times Ahead for Telescope-Makers." *Science* 240 (1 April 1988): 28–31.

Washburn, Mark. "Goodbye, Voyager." *Air & Space Smithsonian* 4 (December 1989/January 1990): 38–48.

Weaver, Kenneth F. "Countdown for Space." *National Geographic Magazine* 119 (May 1961): 702–734.

Webb, George E. "The Chemist as Consultant in Gilded Age America: Benjamin Silliman, Jr. and Western Mining." *Bulletin for the History of Chemistry* 15/16 (1994): 9–14.

——. "Dendrochronology." In *Sciences of the Earth: An Encyclopedia of Events, People, and Phenomena*, edited by Gregory A. Good. New York: Garland Publishing, Inc., 1998.

——. "The Evolution Controversy in Arizona and California: From the 1920s to the 1980s." *Journal of the Southwest* 33 (Summer 1991): 133–150.

——. "The Indefatigable Astronomer: A. E. Douglass and the Founding of the Steward Observatory." *Journal of Arizona History* 19 (Summer 1978): 169–188.

——. "Kitt Peak National Observatory." In *History of Astronomy: An Encyclopedia*, edited by John Lankford. New York: Garland Publishing, Inc., 1997.

——. "Leading Women Scientists in the American Southwest: A Demographic Portrait, 1900–1950." *New Mexico Historical Review* 68 (January 1993): 41–61.

——. "Multiple Mirror Telescopes." In *History of Astronomy: An Encyclopedia*, edited by John Lankford. New York: Garland Publishing, Inc., 1997.

——. "Scientists in the American Southwest: The Birth of a Community, 1906–1938." *The Historian* 50 (February 1988): 173–195.

——. "Solar Physics and the Origins of Dendrochronology." *Isis* 77 (June 1986): 291–301.

——. "Tucson's Evolution Debate, 1924–1927." *Journal of Arizona History* 24 (Spring 1983): 1–12.

——. "University of Arizona, Astronomy at." In *History of Astronomy: An Encyclopedia*, edited by John Lankford. New York: Garland Publishing, Inc., 1997.

——. "A Woman's Place is in the Lab: Arizona's Women Research Scientists, 1910–1950." *Journal of Arizona History* 34 (Spring 1993): 45–64.

——, ed. "Benjamin Silliman's Visit to California: A Letter to His Wife, 1864." *Southern California Quarterly* 59 (Winter 1977): 365–378.

——, ed. "The Mines in Northwestern Arizona in 1864: A Report by Benjamin Silliman, Jr." *Arizona and the West* 16 (Autumn 1974): 247–270.

Wheelwright, Jeff. "For our nuclear wastes, there's gridlock on the road to the dump." *Smithsonian* 26 (May 1995): 40–50.

White, Gerald T. "The Case of the Salted Sample: A California Oil Industry Skeleton." *Pacific Historical Review* 35 (May 1966): 153–184.

Williamson, R. A. "Casa Rinconada, A Twelfth Century Anasazi Kiva." In *Archaeoastronomy in the New World*, edited by Anthony F. Aveni. Cambridge: Cambridge University Press, 1982.

Williamson, R. A., et al. "Anasazi Solar Observatories." In *Native American Astronomy*, edited by Anthony F. Aveni. Austin: University of Texas Press, 1977.

Wilson, Steve. "Mesquite: The Forgotten Manna of the Southwest." *Great Plains Journal* 23 (1984): 83–105.

Wood, Frank Bradshaw. "The Eclipsing Variable RX Herculis." *Astrophysical Journal* 110 (November 1949): 465–474.

Wright, Arthur. "Benjamin Silliman, 1816–1885." National Academy of Sciences, *Biographical Memoirs* 7 (1911): 117–141.

Wuethrich, Bernice. "Scientists Strike Back against Creationism." *Science* 286 (22 October 1999): 659.

Zhu, Liping. "The Historical Statistics of the New Mexico Mining Industry." *Mining History Journal* 2 (1995): 91–98.

Index

About the Author

George E. Webb is a specialist in the history of American science, with a primary interest in the Southwest. A graduate of the University of Arizona (Ph.D. 1978), he has published in scholarly journals including *Isis*, the *Journal of Arizona History*, the *New Mexico Historical Review*, the *Journal of the Southwest*, the *Bulletin for the History of Chemistry*, and *Astronomy Quarterly*. He has presented his research at meetings of the Western History Association, American Astronomical Society, History of Science Society, American Anthropological Association, and the state historical societies of Arizona and New Mexico. His current research includes an investigation of scientists' role in the early development of the Arizona date industry and an examination of the political and cultural aspects of the establishment of the Dominion Astrophysical Observatory in Victoria, British Columbia. Webb is currently a professor of history at Tennessee Technological University, where he teaches courses in the history of science and the history of the American West. Webb's first book, *Tree Rings and Telescopes: The Scientific Career of A. E. Douglass*, was also published by the University of Arizona Press.